POCKET BOOK OF STATISTICAL TABLES

Compiled by

ROBERT E. ODEH
Department of Mathematics
University of Victoria
Victoria, British Columbia
Canada

DONALD B. OWEN
Department of Statistics
Southern Methodist University
Dallas, Texas

Z. W. BIRNBAUM
Department of Mathematics
University of Washington
Seattle, Washington

LLOYD FISHER
Department of Biostatistics
University of Washington
Seattle, Washington

MARCEL DEKKER, INC. New York and Basel

Library of Congress Cataloging in Publication Data

Main entry under title:

Pocket book of statistical tables.

(Statistics, textbooks and monographs ; v. 22)
Includes bibliographical references.
1. Mathematical statistics--Tables, etc.
I. Odeh, Robert E.
QA276.25.P6 519.5'021'2 76-28106
ISBN 0-8247-6515-X

MARCEL DEKKER, INC.
270 Madison Avenue, New York, New York 10016

Current printing (last digit):
10 9 8 7 6 5 4 3 2 1

PRINTED IN THE UNITED STATES OF AMERICA

DEDICATIONS

To my parents on the occasion of their 50th wedding anniversary and to my family - R.E.O.

To my wife and our children - D.B.O.

To Hilde - Z.W.B.

To Ginny, Laura, and Brad - L.F.

INTRODUCTION

The advent of electronic hand calculators has greatly changed the need for and use of mathematical tables. In particular, much of the traditional contents of collections of tables for statisticians has become unnecessary since many useful functions, such as the exponential and trigonometric, and their inverses, powers, logarithms, etc., are now programmed into hand calculators and can be computed more accurately and much faster than one would obtain them from tables. At the same time, the small size of these calculators enables the statistician to perform many calculations conveniently and quickly while away from his office and a desk calculator or computer.

It is the aim of this pocket book to provide a collection of statistical tables which take these developments into account in two ways: it eliminates traditional tables made obsolete by many hand calculators, and it is of a size which matches a pocket calculator in convenience.

All of the tables, except Table 20, have been newly calculated and reproduced directly from computer printout, avoiding the possibility of typesetting errors. Some tables are new, contain corrected values, or have not been included in many collections of statistical tables. Existing tables were used wherever possible to check the computed values.

To keep down the size of the book it was necessary to make the descriptions which accompany the tables quite concise. They are intended for readers who are already familiar with the underlying techniques, but are not elaborate enough to teach these techniques. Also, in order to keep down the size and cost of the book, it was decided to put the descriptions of more than one table on the same page.

References for all the tables are given at the end of this pocket book. The references are not intended to be exhaustive but indicate one or two books or papers that lead to an understanding of the tables as well as other sources that themselves include numerous references. For this reason most of the tables have as one reference the Handbook of Statistical Tables by Donald B. Owen (Addison-Wesley, 1962) since this handbook contains larger amounts of material than given here and includes numerous references.

It is hoped that statisticians will find these tables useful in their daily practice when asked ad hoc questions and while consulting in the field. It is also hoped that the small size will make the tables appropriate for class work and use in conjunction with textbooks.

ACKNOWLEDGMENTS

The tables, except Table 20, were computed at the University of Victoria Computing Center. We wish to express our appreciation to Amanda Nemec and Bruce Wilson who assisted with the computation, formatting, and checking of these tables. Financial support for this project was provided by grants from the University of Victoria, the Government of British Columbia, the Canadian National Research Council, and the Defense Research Board of Canada.

CONTENTS

POCKET BOOK OF STATISTICAL TABLES

Table 1 INVERSE OF THE STANDARDIZED NORMAL DISTRIBUTION (Table: p. 3)

Throughout these tables $N(\mu, \sigma^2)$ denotes the probability distribution of a normal random variable that has mean μ and variance σ^2, hence the distribution function

$$\Phi(x;\ \mu,\ \sigma^2) = \int_{-\infty}^{x} \phi(y;\ \mu,\ \sigma^2)\ dy$$

and

$$\phi(y;\ \mu,\ \sigma^2) = \frac{\exp[-(y - \mu)^2/2\sigma^2]}{\sigma\sqrt{2\pi}}$$

Table 1 contains triples of numbers (P, X, Z) such that

$$\Phi(X;\ 0,\ 1) = P, \quad \phi(X;\ 0,\ 1) = Z$$

Table 2 CRITICAL VALUES FOR STUDENT'S t (Table: pp. 4-5)

Table 2 contains solutions t of the equation

$$\frac{\Gamma\{(f + 1)/2\}}{\sqrt{\pi f}\ \Gamma(f/2)} \int_{-\infty}^{t} \left(1 + \frac{x^2}{f}\right)^{-(f+1)/2} dx = \gamma$$

for values of f (degrees of freedom), given in the first column, and values of γ (the cumulative probability), given in the top row.

Table 1 INVERSE OF THE STANDARDIZED NORMAL DISTRIBUTION (Description: p. 2)

P	X	Z	P	X	Z
0.50	0.00000	0.39894	0.925	1.43953	0.14156
0.51	0.02507	0.39882	0.930	1.47579	0.13427
0.52	0.05015	0.39844	0.935	1.51410	0.12679
0.53	0.07527	0.39781	0.940	1.55477	0.11912
0.54	0.10043	0.39694	0.945	1.59819	0.11124
0.55	0.12566	0.39580	0.950	1.64485	0.10314
0.56	0.15097	0.39442	0.955	1.69540	0.09479
0.57	0.17637	0.39279	0.960	1.75069	0.08617
0.58	0.20189	0.39089	0.965	1.81191	0.07727
0.59	0.22754	0.38875	0.970	1.88079	0.06804
0.60	0.25335	0.38634	0.975	1.95996	0.05845
0.61	0.27932	0.38368	0.980	2.05375	0.04842
0.62	0.30548	0.38076	0.985	2.17009	0.03787
0.63	0.33185	0.37757	0.990	2.32635	0.02665
0.64	0.35846	0.37412			
0.65	0.38532	0.37040	0.991	2.36562	0.02431
0.66	0.41246	0.36641	0.992	2.40892	0.02192
0.67	0.43991	0.36215	0.993	2.45726	0.01949
0.68	0.46770	0.35761	0.994	2.51214	0.01700
0.69	0.49585	0.35279	0.995	2.57583	0.01446
0.70	0.52440	0.34769	0.996	2.65207	0.01185
0.71	0.55338	0.34230	0.997	2.74778	0.00915
0.72	0.58284	0.33662	0.998	2.87816	0.00634
0.73	0.61281	0.33065	0.999	3.09023	0.00337
0.74	0.64335	0.32437			
0.75	0.67449	0.31778	0.9991	3.12139	0.00306
0.76	0.70630	0.31087	0.9992	3.15591	0.00274
0.77	0.73885	0.30365	0.9993	3.19465	0.00243
0.78	0.77219	0.29609	0.9994	3.23888	0.00210
0.79	0.80642	0.28820	0.9995	3.29053	0.00178
0.80	0.84162	0.27996	0.9996	3.35279	0.00145
0.81	0.87790	0.27137	0.9997	3.43161	0.00111
0.82	0.91537	0.26240	0.9998	3.54008	0.00076
0.83	0.95417	0.25305	0.9999	3.71902	0.00040
0.84	0.99446	0.24331			
0.85	1.03643	0.23316	0.9999 5	3.89059	0.00021
0.86	1.08032	0.22258	0.9999 9	4.26489	0.00004
0.87	1.12639	0.21155	0.9999 95	4.41717	0.00002
0.88	1.17499	0.20004	0.9999 99	4.75342	0.00000
0.89	1.22653	0.18804	0.9999 995	4.89164	0.00000
0.900	1.28155	0.17550	0.9999 999	5.19934	0.00000
0.905	1.31058	0.16902	0.9999 9995	5.32672	0.00000
0.910	1.34076	0.16239	0.9999 9999	5.61200	0.00000
0.915	1.37220	0.15561	0.9999 9999 5	5.73073	0.00000
0.920	1.40507	0.14867	0.9999 9999 9	5.99781	0.00000

Table 2 CRITICAL VALUES FOR STUDENT'S t
(Description: p. 2)

	cumulative probability, γ						
f	0.70	0.80	0.90	0.95	0.975	0.990	0.995
1	0.727	1.376	3.0777	6.3138	12.7062	31.8205	63.6567
2	0.617	1.061	1.8856	2.9200	4.3027	6.9646	9.9248
3	0.584	0.978	1.6378	2.3534	3.1824	4.5407	5.8409
4	0.569	0.941	1.5332	2.1319	2.7766	3.7470	4.6041
5	0.559	0.920	1.4759	2.0151	2.5706	3.3651	4.0322
6	0.553	0.906	1.4398	1.9432	2.4469	3.1427	3.7075
7	0.549	0.896	1.4149	1.8946	2.3646	2.9980	3.4995
8	0.546	0.889	1.3968	1.8595	2.3060	2.8965	3.3554
9	0.543	0.883	1.3830	1.8331	2.2622	2.8214	3.2498
10	0.542	0.879	1.3722	1.8125	2.2281	2.7638	3.1693
11	0.540	0.876	1.3634	1.7959	2.2010	2.7181	3.1058
12	0.539	0.873	1.3562	1.7823	2.1788	2.6810	3.0545
13	0.538	0.870	1.3502	1.7709	2.1604	2.6503	3.0123
14	0.537	0.868	1.3450	1.7613	2.1448	2.6245	2.9768
15	0.536	0.866	1.3406	1.7531	2.1314	2.6025	2.9467
16	0.535	0.865	1.3368	1.7459	2.1199	2.5835	2.9208
17	0.534	0.863	1.3334	1.7396	2.1098	2.5669	2.8982
18	0.534	0.862	1.3304	1.7341	2.1009	2.5524	2.8784
19	0.533	0.861	1.3277	1.7291	2.0930	2.5395	2.8609
20	0.533	0.860	1.3253	1.7247	2.0860	2.5280	2.8453
21	0.532	0.859	1.3232	1.7207	2.0796	2.5176	2.8314
22	0.532	0.858	1.3212	1.7171	2.0739	2.5083	2.8188
23	0.532	0.858	1.3195	1.7139	2.0687	2.4999	2.8073
24	0.531	0.857	1.3178	1.7109	2.0639	2.4922	2.7969
25	0.531	0.856	1.3163	1.7081	2.0595	2.4851	2.7874
26	0.531	0.856	1.3150	1.7056	2.0555	2.4786	2.7787
27	0.531	0.855	1.3137	1.7033	2.0518	2.4727	2.7707
28	0.530	0.855	1.3125	1.7011	2.0484	2.4671	2.7633
29	0.530	0.854	1.3114	1.6991	2.0452	2.4620	2.7564
30	0.530	0.854	1.3104	1.6973	2.0423	2.4573	2.7500
31	0.530	0.853	1.3095	1.6955	2.0395	2.4528	2.7440
32	0.530	0.853	1.3086	1.6939	2.0369	2.4487	2.7385
33	0.530	0.853	1.3077	1.6924	2.0345	2.4448	2.7333
34	0.529	0.852	1.3070	1.6909	2.0322	2.4411	2.7284
35	0.529	0.852	1.3062	1.6896	2.0301	2.4377	2.7238
36	0.529	0.852	1.3055	1.6883	2.0281	2.4345	2.7195
37	0.529	0.851	1.3049	1.6871	2.0262	2.4314	2.7154
38	0.529	0.851	1.3042	1.6860	2.0244	2.4286	2.7116
39	0.529	0.851	1.3036	1.6849	2.0227	2.4258	2.7079
40	0.529	0.851	1.3031	1.6839	2.0211	2.4233	2.7045

Table 2 CRITICAL VALUES FOR STUDENT'S t
(Description: p. 2)

f	cumulative probability, γ						
	0.70	0.80	0.90	0.95	0.975	0.990	0.995
42	0.528	0.850	1.3020	1.6820	2.0181	2.4185	2.6981
44	0.528	0.850	1.3011	1.6802	2.0154	2.4141	2.6923
46	0.528	0.850	1.3002	1.6787	2.0129	2.4102	2.6870
48	0.528	0.849	1.2994	1.6772	2.0106	2.4066	2.6822
50	0.528	0.849	1.2987	1.6759	2.0086	2.4033	2.6778
55	0.527	0.848	1.2971	1.6730	2.0040	2.3961	2.6682
60	0.527	0.848	1.2958	1.6706	2.0003	2.3901	2.6603
65	0.527	0.847	1.2947	1.6686	1.9971	2.3851	2.6536
70	0.527	0.847	1.2938	1.6669	1.9944	2.3808	2.6479
75	0.527	0.846	1.2929	1.6654	1.9921	2.3771	2.6430
80	0.526	0.846	1.2922	1.6641	1.9901	2.3739	2.6387
85	0.526	0.846	1.2916	1.6630	1.9883	2.3710	2.6349
90	0.526	0.846	1.2910	1.6620	1.9867	2.3685	2.6316
95	0.526	0.845	1.2905	1.6611	1.9853	2.3662	2.6286
100	0.526	0.845	1.2901	1.6602	1.9840	2.3642	2.6259
110	0.526	0.845	1.2893	1.6588	1.9818	2.3607	2.6213
120	0.526	0.845	1.2886	1.6577	1.9799	2.3578	2.6174
130	0.526	0.844	1.2881	1.6567	1.9784	2.3554	2.6142
140	0.526	0.844	1.2876	1.6558	1.9771	2.3533	2.6114
150	0.526	0.844	1.2872	1.6551	1.9759	2.3515	2.6090
160	0.525	0.844	1.2869	1.6544	1.9749	2.3499	2.6069
170	0.525	0.844	1.2866	1.6539	1.9740	2.3485	2.6051
180	0.525	0.844	1.2863	1.6534	1.9732	2.3472	2.6034
190	0.525	0.844	1.2860	1.6529	1.9725	2.3461	2.6020
200	0.525	0.843	1.2858	1.6525	1.9719	2.3451	2.6006
300	0.525	0.843	1.2844	1.6499	1.9679	2.3388	2.5923
400	0.525	0.843	1.2837	1.6487	1.9659	2.3357	2.5882
500	0.525	0.842	1.2832	1.6479	1.9647	2.3338	2.5857
600	0.525	0.842	1.2830	1.6474	1.9639	2.3326	2.5840
700	0.525	0.842	1.2828	1.6470	1.9634	2.3317	2.5829
800	0.525	0.842	1.2826	1.6468	1.9629	2.3310	2.5820
900	0.525	0.842	1.2825	1.6465	1.9626	2.3305	2.5813
1000	0.525	0.842	1.2824	1.6464	1.9623	2.3301	2.5808
∞	0.524	0.842	1.2816	1.6449	1.9600	2.3263	2.5758

__Table 3__ SAMPLE SIZE FOR A TEST OF THE MEAN OF A NORMAL VARIABLE (Table: pp. 7-8)

A one-sample t-test is to be used to test that the mean of a normal population is equal to a hypothesized value, μ_0. Let the true mean of the population be μ and the standard deviation be σ. When testing at a given significance level α and when the true mean is μ, it is desired that the probability of accepting the hypothesis $\mu = \mu_0$ be less than or equal to β. The table entry is the minimum sample size needed to attain the type II error, β, for given α, β and $\Delta = |\mu - \mu_0|/\sigma$.

__Table 4__ SAMPLE SIZE FOR COMPARING THE MEANS OF TWO NORMAL VARIABLES (Table: pp. 9-10)

The two-sample t-test for two normal populations with common variance and possibly unequal means is to be used. Let $\Delta = |\mu_1 - \mu_2|/\sigma$ be the true absolute difference between the means in common standard deviation units. For given α (the significance level), Δ and β (the probability of a type II error), the table entry gives the __minimum sample size needed for each group__ assuming that the two groups are to be sampled an equal number of times.

Table [illegible] SAMPLE SIZE FOR A TEST OF THE MEAN OF A NORMAL VARIABLE (Description: p. 6)

α = 0.05 (one-sided), 0.10 (two-sided)

Δ	β, probability of type II error								
	0.80	0.70	0.60	0.50	0.40	0.30	0.20	0.10	0.05
0.3	13	23	23	33	43	63	73	103	124
0.4	6	10	14	19	24	31	41	55	70
0.5	5	7	10	13	16	21	27	36	45
0.6	4	6	7	9	12	15	19	20	32
0.7	3	5	6	8	9	12	15	19	24
0.8	3	4	5	6	8	9	12	15	19
0.9	3	4	4	5	7	8	10	13	15
1.0	3	3	4	5	6	7	8	11	13
1.2	3	3	3	4	5	5	6	8	10
1.4	3	3	3	4	4	5	5	7	8
1.6	3	3	3	3	4	4	5	6	6
1.8	3	3	3	3	3	4	4	5	6
2.0	3	3	3	3	3	3	4	4	5
3.0	3	3	3	3	3	3	3	3	4
4.0	3	3	3	3	3	3	3	3	3

α = 0.025 (one-sided), 0.05 (two-sided)

Δ	β, probability of type II error								
	0.80	0.70	0.60	0.50	0.40	0.30	0.20	0.10	0.05
0.3	23	33	43	53	63	73	94	124	154
0.4	10	15	21	26	33	41	52	68	84
0.5	7	11	14	18	22	27	34	44	54
0.6	6	8	11	13	16	20	24	32	39
0.7	5	7	8	10	13	15	19	24	29
0.8	4	6	7	9	10	12	15	19	23
0.9	4	5	6	7	9	10	12	16	19
1.0	4	5	5	6	7	9	10	13	16
1.2	3	4	5	5	6	7	8	10	12
1.4	3	4	4	5	5	6	7	8	9
1.6	3	3	4	4	5	5	6	7	8
1.8	3	3	3	4	4	5	5	6	7
2.0	3	3	3	4	4	4	5	5	6
3.0	3	3	3	3	3	3	4	4	4
4.0	3	3	3	3	3	3	3	3	4

Table 3 SAMPLE SIZE FOR A TEST OF THE MEAN OF A NORMAL VARIABLE (Description: p. 6)

α = 0.01 (one-sided), 0.02 (two-sided)

	β, probability of type II error								
Δ	0.90	0.80	0.70	0.60	0.50	0.40	0.30	0.20	0.10
0.3	23	33	43	54	64	84	94	124	155
0.4	10	17	24	30	37	45	54	66	85
0.5	7	12	16	20	25	30	36	43	55
0.6	6	9	12	15	18	22	26	31	39
0.7	5	8	10	12	14	17	20	24	30
0.8	5	7	8	10	12	14	16	19	24
0.9	4	6	7	9	10	12	13	16	19
1.0	4	5	6	8	9	10	11	13	16
1.2	3	5	5	6	7	8	9	10	12
1.4	3	4	5	5	6	7	7	8	10
1.6	3	4	4	5	5	6	7	7	9
1.8	3	4	4	4	5	5	6	6	7
2.0	3	3	4	4	5	5	5	6	7
3.0	3	3	3	3	4	4	4	4	5
4.0	3	3	3	3	3	3	4	4	4

α = 0.005 (one-sided), 0.01 (two-sided)

	β, probability of type II error								
Δ	0.90	0.80	0.70	0.60	0.50	0.40	0.30	0.20	0.10
0.3	22	37	51	64	80	100	120	140	170
0.4	14	23	30	38	45	54	64	77	97
0.5	10	16	21	25	30	36	42	51	63
0.6	8	12	15	19	22	26	31	36	45
0.7	7	10	12	15	17	20	23	28	34
0.8	6	8	10	12	14	16	19	22	27
0.9	5	7	9	10	12	14	16	18	22
1.0	5	7	8	9	10	12	13	16	19
1.2	4	6	7	7	8	9	11	12	14
1.4	4	5	6	6	7	8	9	10	12
1.6	4	5	5	6	6	7	8	8	10
1.8	3	4	5	5	6	6	7	8	9
2.0	3	4	4	5	5	6	6	7	8
3.0	3	3	4	4	4	4	5	5	6
4.0	3	3	3	4	4	4	4	4	5

Table 4 SAMPLE SIZE FOR COMPARING THE MEANS OF TWO NORMAL VARIABLES (Description: p. 6)

α = 0.05 (one-sided), 0.10 (two-sided)

	β, probability of type II error								
Δ	0.80	0.70	0.60	0.50	0.40	0.30	0.20	0.10	0.05
0.3	16	29	44	61	81	106	139	191	242
0.4	9	17	25	35	46	60	78	108	136
0.5	6	11	17	23	30	39	51	70	88
0.6	5	8	12	16	21	27	36	49	61
0.7	4	6	9	12	16	20	26	36	45
0.8	3	5	7	10	12	16	21	28	35
0.9	3	4	6	8	10	13	16	22	28
1.0	3	4	5	7	8	11	14	18	23
1.2	2	3	4	5	6	8	10	13	16
1.4	2	3	3	4	5	6	8	10	12
1.6	2	3	3	4	4	5	6	8	10
1.8	2	2	3	3	4	4	5	7	8
2.0	2	2	3	3	3	4	4	6	7
3.0	2	2	2	2	2	3	3	3	4
4.0	2	2	2	2	2	2	2	3	3

α = 0.025 (one-sided), 0.05 (two-sided)

	β, probability of type II error								
Δ	0.80	0.70	0.60	0.50	0.40	0.30	0.20	0.10	0.05
0.3	29	47	66	87	110	139	176	235	290
0.4	17	27	38	49	63	79	100	133	164
0.5	12	18	25	32	41	51	64	86	105
0.6	8	13	18	23	29	36	45	60	74
0.7	7	10	13	17	21	27	34	44	55
0.8	6	8	11	14	17	21	26	34	42
0.9	5	7	9	11	14	17	21	27	34
1.0	4	6	7	9	11	14	17	23	27
1.2	3	5	6	7	8	10	12	16	20
1.4	3	4	5	6	7	8	10	12	15
1.6	3	3	4	5	5	6	8	10	12
1.8	3	3	4	4	5	5	6	8	10
2.0	2	3	3	4	4	5	6	7	8
3.0	2	2	3	3	3	3	4	4	5
4.0	2	2	2	2	3	3	3	3	4

Table 4 SAMPLE SIZE FOR COMPARING THE MEANS OF TWO NORMAL VARIABLES (Description: p. 6)

α = 0.01 (one-sided), 0.02 (two-sided)

Δ	β, probability of type II error								
	0.90	0.80	0.70	0.60	0.50	0.40	0.30	0.20	0.10
0.3	26	51	74	97	122	150	182	225	291
0.4	16	29	42	56	70	85	103	127	165
0.5	11	20	28	36	45	55	67	82	106
0.6	8	14	20	26	32	39	47	58	74
0.7	6	11	15	19	24	29	35	43	55
0.8	5	9	12	15	19	23	27	33	43
0.9	5	7	10	13	15	18	22	27	34
1.0	4	6	8	11	13	15	18	22	28
1.2	4	5	6	8	9	11	13	16	20
1.4	3	4	5	6	8	9	10	12	15
1.6	3	4	5	5	6	7	8	10	12
1.8	3	3	4	5	5	6	7	8	10
2.0	3	3	4	4	5	5	6	7	9
3.0	2	3	3	3	3	4	4	4	5
4.0	2	2	3	3	3	3	3	3	4

α = 0.005 (one-sided), 0.01 (two-sided)

Δ	β, probability of type II error								
	0.90	0.80	0.70	0.60	0.50	0.40	0.30	0.20	0.10
0.3	39	69	96	122	150	180	216	262	333
0.4	23	40	55	70	85	102	122	148	188
0.5	16	26	36	45	55	66	79	96	121
0.6	12	19	26	32	39	47	56	67	85
0.7	9	14	19	24	29	35	41	50	63
0.8	7	12	15	19	23	27	32	39	49
0.9	6	10	13	16	19	22	26	31	39
1.0	6	8	11	13	15	18	21	26	32
1.2	5	6	8	10	11	13	16	18	23
1.4	4	5	7	8	9	10	12	14	17
1.6	4	5	6	7	7	9	10	11	14
1.8	3	4	5	6	6	7	8	10	11
2.0	3	4	4	5	6	6	7	8	10
3.0	3	3	3	4	4	4	5	5	6
4.0	2	3	3	3	3	3	4	4	4

Table 5 CRITICAL VALUES OF THE CHI SQUARE DISTRIBUTION
(Table: pp. 12-13)

Table 5 contains values obtained by solving for u the equation

$$\frac{1}{2^{f/2}\,\Gamma(f/2)}\int_0^u x^{(f-2)/2}\,e^{-x/2}\,dx = \gamma$$

for values f (the degrees of freedom), given in the first column, and values of γ (the cumulative probability), given in the top row.

Table 6 CRITICAL VALUES OF THE F-DISTRIBUTION
(Table: pp. 14-19)

This table presents the upper 10%, 5%, and 1% critical values for the F-distribution with f_1 (the numerator degrees of freedom) and f_2 (the denominator degrees of freedom). That is, for α = .10, .05, and .01 and tabulated values of f_1 (column heading) and f_2 (row heading), the value F is presented for which

$$\int_F^{\infty} \frac{\Gamma(\{f_1 + f_2\}/2)\,(f_1/f_2)^{f_1/2}\,x^{(f_1-2)/2}\,dx}{\Gamma(f_1/2)\Gamma(f_2/2)\,(1 + \{f_1/f_2\}x)^{(f_1+f_2)/2}} = \alpha$$

Table 5 CRITICAL VALUES OF THE CHI SQUARE DISTRIBUTION (Description: p. 11)

f	cumulative probability, γ					
	0.005	0.010	0.025	0.050	0.100	0.250
1	-	-	0.001	0.004	0.016	0.102
2	0.010	0.020	0.051	0.103	0.211	0.575
3	0.072	0.115	0.216	0.352	0.584	1.213
4	0.207	0.297	0.484	0.711	1.064	1.923
5	0.412	0.554	0.831	1.145	1.610	2.675
6	0.676	0.872	1.237	1.635	2.204	3.455
7	0.989	1.239	1.690	2.167	2.833	4.255
8	1.344	1.646	2.180	2.733	3.490	5.071
9	1.735	2.088	2.700	3.325	4.168	5.899
10	2.156	2.558	3.247	3.940	4.865	6.737
11	2.603	3.053	3.816	4.575	5.578	7.584
12	3.074	3.571	4.404	5.226	6.304	8.438
13	3.565	4.107	5.009	5.892	7.042	9.299
14	4.075	4.660	5.629	6.571	7.790	10.165
15	4.601	5.229	6.262	7.261	8.547	11.037
16	5.142	5.812	6.908	7.962	9.312	11.912
17	5.697	6.408	7.564	8.672	10.085	12.792
18	6.265	7.015	8.231	9.390	10.865	13.675
19	6.844	7.633	8.907	10.117	11.651	14.562
20	7.434	8.260	9.591	10.851	12.443	15.452
21	8.034	8.897	10.283	11.591	13.240	16.344
22	8.643	9.542	10.982	12.338	14.041	17.240
23	9.260	10.196	11.689	13.091	14.848	18.137
24	9.886	10.856	12.401	13.848	15.659	19.037
25	10.520	11.524	13.120	14.611	16.473	19.939
26	11.160	12.198	13.844	15.379	17.292	20.843
27	11.808	12.879	14.573	16.151	18.114	21.749
28	12.461	13.565	15.308	16.928	18.939	22.657
29	13.121	14.256	16.047	17.708	19.768	23.567
30	13.787	14.953	16.791	18.493	20.599	24.478
35	17.192	18.509	20.569	22.465	24.797	29.054
40	20.707	22.164	24.433	26.509	29.051	33.660
45	24.311	25.901	28.366	30.612	33.350	38.291
50	27.991	29.707	32.357	34.764	37.689	42.942
55	31.735	33.570	36.398	38.958	42.060	47.610
60	35.534	37.485	40.482	43.188	46.459	52.294
65	39.383	41.444	44.603	47.450	50.883	56.990
70	43.275	45.442	48.758	51.739	55.329	61.698
75	47.206	49.475	52.942	56.054	59.795	66.417
80	51.172	53.540	57.153	60.391	64.278	71.145
85	55.170	57.634	61.389	64.749	68.777	75.881
90	59.196	61.754	65.647	69.126	73.291	80.625
95	63.250	65.898	69.925	73.520	77.818	85.376
100	67.328	70.065	74.222	77.929	82.358	90.133
120	83.852	86.923	91.573	95.705	100.624	109.220

Table 5 CRITICAL VALUES OF THE CHI SQUARE DISTRIBUTION
(Description: p. 11)

f	cumulative probability, γ					
	0.750	0.900	0.950	0.975	0.990	0.995
1	1.323	2.706	3.841	5.024	6.635	7.879
2	2.773	4.605	5.991	7.378	9.210	10.597
3	4.108	6.251	7.815	9.348	11.345	12.838
4	5.385	7.779	9.488	11.143	13.277	14.860
5	6.626	9.236	11.070	12.833	15.086	16.750
6	7.841	10.645	12.592	14.449	16.812	18.548
7	9.037	12.017	14.067	16.013	18.475	20.278
8	10.219	13.362	15.507	17.535	20.090	21.955
9	11.389	14.684	16.919	19.023	21.666	23.589
10	12.549	15.987	18.307	20.483	23.209	25.188
11	13.701	17.275	19.675	21.920	24.725	26.757
12	14.845	18.549	21.026	23.337	26.217	28.300
13	15.984	19.812	22.362	24.736	27.688	29.819
14	17.117	21.064	23.685	26.119	29.141	31.319
15	18.245	22.307	24.996	27.488	30.578	32.801
16	19.369	23.542	26.296	28.845	32.000	34.267
17	20.489	24.769	27.587	30.191	33.409	35.718
18	21.605	25.989	28.869	31.526	34.805	37.156
19	22.718	27.204	30.144	32.852	36.191	38.582
20	23.828	28.412	31.410	34.170	37.566	39.997
21	24.935	29.615	32.671	35.479	38.932	41.401
22	26.039	30.813	33.924	36.781	40.289	42.796
23	27.141	32.007	35.172	38.076	41.638	44.181
24	28.241	33.196	36.415	39.364	42.980	45.559
25	29.339	34.382	37.652	40.646	44.314	46.928
26	30.435	35.563	38.885	41.923	45.642	48.290
27	31.528	36.741	40.113	43.195	46.963	49.645
28	32.620	37.916	41.337	44.461	48.278	50.993
29	33.711	39.087	42.557	45.722	49.588	52.336
30	34.800	40.256	43.773	46.979	50.892	53.672
35	40.223	46.059	49.802	53.203	57.342	60.275
40	45.616	51.805	55.758	59.342	63.691	66.766
45	50.985	57.505	61.656	65.410	69.957	73.166
50	56.334	63.167	67.505	71.420	76.154	79.490
55	61.665	68.796	73.311	77.380	82.292	85.749
60	66.981	74.397	79.082	83.298	88.379	91.952
65	72.205	79.973	84.821	89.177	94.422	98.105
70	77.577	85.527	90.531	95.023	100.425	104.215
75	82.858	91.061	96.217	100.839	106.393	110.286
80	88.130	96.578	101.879	106.629	112.329	116.321
85	93.394	102.079	107.522	112.393	118.236	122.325
90	98.650	107.565	113.145	118.136	124.116	128.299
95	103.899	113.038	118.752	123.858	129.973	134.247
100	109.141	118.498	124.342	129.561	135.807	140.169
120	130.055	140.233	146.567	152.211	158.950	163.648

Table 6 CRITICAL VALUES OF THE F-DISTRIBUTION
(Description: p. 11)

$\alpha = 0.10$

$f_2 \backslash f_1$	1	2	3	4	5	6	7	8	9
2	8.53	9.00	9.16	9.24	9.29	9.33	9.35	9.37	9.38
3	5.54	5.46	5.39	5.34	5.31	5.28	5.27	5.25	5.24
4	4.54	4.32	4.19	4.11	4.05	4.01	3.98	3.95	3.94
5	4.06	3.78	3.62	3.52	3.45	3.40	3.37	3.34	3.32
6	3.78	3.46	3.29	3.18	3.11	3.05	3.01	2.98	2.96
7	3.59	3.26	3.07	2.96	2.88	2.83	2.78	2.75	2.72
8	3.46	3.11	2.92	2.81	2.73	2.67	2.62	2.59	2.56
9	3.36	3.01	2.81	2.69	2.61	2.55	2.51	2.47	2.44
10	3.29	2.92	2.73	2.61	2.52	2.46	2.41	2.38	2.35
11	3.23	2.86	2.66	2.54	2.45	2.39	2.34	2.30	2.27
12	3.18	2.81	2.61	2.48	2.39	2.33	2.28	2.24	2.21
13	3.14	2.76	2.56	2.43	2.35	2.28	2.23	2.20	2.16
14	3.10	2.73	2.52	2.39	2.31	2.24	2.19	2.15	2.12
15	3.07	2.70	2.49	2.36	2.27	2.21	2.16	2.12	2.09
16	3.05	2.67	2.46	2.33	2.24	2.18	2.13	2.09	2.06
17	3.03	2.64	2.44	2.31	2.22	2.15	2.10	2.06	2.03
18	3.01	2.62	2.42	2.29	2.20	2.13	2.08	2.04	2.00
19	2.99	2.61	2.40	2.27	2.18	2.11	2.06	2.02	1.98
20	2.97	2.59	2.38	2.25	2.16	2.09	2.04	2.00	1.96
21	2.96	2.57	2.36	2.23	2.14	2.08	2.02	1.98	1.95
22	2.95	2.56	2.35	2.22	2.13	2.06	2.01	1.97	1.93
23	2.94	2.55	2.34	2.21	2.11	2.05	1.99	1.95	1.92
24	2.93	2.54	2.33	2.19	2.10	2.04	1.98	1.94	1.91
25	2.92	2.53	2.32	2.18	2.09	2.02	1.97	1.93	1.89
26	2.91	2.52	2.31	2.17	2.08	2.01	1.96	1.92	1.88
27	2.90	2.51	2.30	2.17	2.07	2.00	1.95	1.91	1.87
28	2.89	2.50	2.29	2.16	2.06	2.00	1.94	1.90	1.87
29	2.89	2.50	2.28	2.15	2.06	1.99	1.93	1.89	1.86
30	2.88	2.49	2.28	2.14	2.05	1.98	1.93	1.88	1.85
40	2.84	2.44	2.23	2.09	2.00	1.93	1.87	1.83	1.79
48	2.81	2.42	2.20	2.07	1.97	1.90	1.85	1.80	1.77
60	2.79	2.39	2.18	2.04	1.95	1.87	1.82	1.77	1.74
90	2.76	2.36	2.15	2.01	1.91	1.84	1.78	1.74	1.70
120	2.75	2.35	2.13	1.99	1.90	1.82	1.77	1.72	1.68
∞	2.71	2.30	2.08	1.94	1.85	1.77	1.72	1.67	1.63

Table 6 CRITICAL VALUES OF THE F-DISTRIBUTION
(Description: p. 11)

$\alpha = 0.10$

$f_2 \backslash f_1$	10	12	15	20	24	30	40	60	120
2	9.39	9.41	9.42	9.44	9.45	9.46	9.47	9.47	9.48
3	5.23	5.22	5.20	5.18	5.18	5.17	5.16	5.15	5.14
4	3.92	3.90	3.87	3.84	3.83	3.82	3.80	3.79	3.78
5	3.30	3.27	3.24	3.21	3.19	3.17	3.16	3.14	3.12
6	2.94	2.90	2.87	2.84	2.82	2.80	2.78	2.76	2.74
7	2.70	2.67	2.63	2.59	2.58	2.56	2.54	2.51	2.49
8	2.54	2.50	2.46	2.42	2.40	2.38	2.36	2.34	2.32
9	2.42	2.38	2.34	2.30	2.28	2.25	2.23	2.21	2.18
10	2.32	2.28	2.24	2.20	2.18	2.16	2.13	2.11	2.08
11	2.25	2.21	2.17	2.12	2.10	2.08	2.05	2.03	2.00
12	2.19	2.15	2.10	2.06	2.04	2.01	1.99	1.96	1.93
13	2.14	2.10	2.05	2.01	1.98	1.96	1.93	1.90	1.88
14	2.10	2.05	2.01	1.96	1.94	1.91	1.89	1.86	1.83
15	2.06	2.02	1.97	1.92	1.90	1.87	1.85	1.82	1.79
16	2.03	1.99	1.94	1.89	1.87	1.84	1.81	1.78	1.75
17	2.00	1.96	1.91	1.86	1.84	1.81	1.78	1.75	1.72
18	1.98	1.93	1.89	1.84	1.81	1.78	1.75	1.72	1.69
19	1.96	1.91	1.86	1.81	1.79	1.76	1.73	1.70	1.67
20	1.94	1.89	1.84	1.79	1.77	1.74	1.71	1.68	1.64
21	1.92	1.87	1.83	1.78	1.75	1.72	1.69	1.66	1.62
22	1.90	1.86	1.81	1.76	1.73	1.70	1.67	1.64	1.60
23	1.89	1.84	1.80	1.74	1.72	1.69	1.66	1.62	1.59
24	1.88	1.83	1.78	1.73	1.70	1.67	1.64	1.61	1.57
25	1.87	1.82	1.77	1.72	1.69	1.66	1.63	1.59	1.56
26	1.86	1.81	1.76	1.71	1.68	1.65	1.61	1.58	1.54
27	1.85	1.80	1.75	1.70	1.67	1.64	1.60	1.57	1.53
28	1.84	1.79	1.74	1.69	1.66	1.63	1.59	1.56	1.52
29	1.83	1.78	1.73	1.68	1.65	1.62	1.58	1.55	1.51
30	1.82	1.77	1.72	1.67	1.64	1.61	1.57	1.54	1.50
40	1.76	1.71	1.66	1.61	1.57	1.54	1.51	1.47	1.42
48	1.73	1.69	1.63	1.57	1.54	1.51	1.47	1.43	1.39
60	1.71	1.66	1.60	1.54	1.51	1.48	1.44	1.40	1.35
90	1.67	1.62	1.56	1.50	1.47	1.43	1.39	1.35	1.29
120	1.65	1.60	1.55	1.48	1.45	1.41	1.37	1.32	1.26
∞	1.60	1.55	1.49	1.42	1.38	1.34	1.30	1.24	1.17

Table 6 CRITICAL VALUES OF THE F-DISTRIBUTION
(Description: p. 11)

$\alpha = 0.05$

f_2 \ f_1	1	2	3	4	5	6	7	8	9
2	18.51	19.00	19.16	19.25	19.30	19.33	19.35	19.37	19.38
3	10.13	9.55	9.28	9.12	9.01	8.94	8.89	8.85	8.81
4	7.71	6.94	6.59	6.39	6.26	6.16	6.09	6.04	6.00
5	6.61	5.79	5.41	5.19	5.05	4.95	4.88	4.82	4.77
6	5.99	5.14	4.76	4.53	4.39	4.28	4.21	4.15	4.10
7	5.59	4.74	4.35	4.12	3.97	3.87	3.79	3.73	3.68
8	5.32	4.46	4.07	3.84	3.69	3.58	3.50	3.44	3.39
9	5.12	4.26	3.86	3.63	3.48	3.37	3.29	3.23	3.18
10	4.96	4.10	3.71	3.48	3.33	3.22	3.14	3.07	3.02
11	4.84	3.98	3.59	3.36	3.20	3.09	3.01	2.95	2.90
12	4.75	3.89	3.49	3.26	3.11	3.00	2.91	2.85	2.80
13	4.67	3.81	3.41	3.18	3.03	2.92	2.83	2.77	2.71
14	4.60	3.74	3.34	3.11	2.96	2.85	2.76	2.70	2.65
15	4.54	3.68	3.29	3.06	2.90	2.79	2.71	2.64	2.59
16	4.49	3.63	3.24	3.01	2.85	2.74	2.66	2.59	2.54
17	4.45	3.59	3.20	2.96	2.81	2.70	2.61	2.55	2.49
18	4.41	3.55	3.16	2.93	2.77	2.66	2.58	2.51	2.46
19	4.38	3.52	3.13	2.90	2.74	2.63	2.54	2.48	2.42
20	4.35	3.49	3.10	2.87	2.71	2.60	2.51	2.45	2.39
21	4.32	3.47	3.07	2.84	2.68	2.57	2.49	2.42	2.37
22	4.30	3.44	3.05	2.82	2.66	2.55	2.46	2.40	2.34
23	4.28	3.42	3.03	2.80	2.64	2.53	2.44	2.37	2.32
24	4.26	3.40	3.01	2.78	2.62	2.51	2.42	2.36	2.30
25	4.24	3.39	2.99	2.76	2.60	2.49	2.40	2.34	2.28
26	4.23	3.37	2.98	2.74	2.59	2.47	2.39	2.32	2.27
27	4.21	3.35	2.96	2.73	2.57	2.46	2.37	2.31	2.25
28	4.20	3.34	2.95	2.71	2.56	2.45	2.36	2.29	2.24
29	4.18	3.33	2.93	2.70	2.55	2.43	2.35	2.28	2.22
30	4.17	3.32	2.92	2.69	2.53	2.42	2.33	2.27	2.21
40	4.08	3.23	2.84	2.61	2.45	2.34	2.25	2.18	2.12
48	4.04	3.19	2.80	2.57	2.41	2.29	2.21	2.14	2.08
60	4.00	3.15	2.76	2.53	2.37	2.25	2.17	2.10	2.04
90	3.95	3.10	2.71	2.47	2.32	2.20	2.11	2.04	1.99
120	3.92	3.07	2.68	2.45	2.29	2.18	2.09	2.02	1.96
∞	3.84	3.00	2.60	2.37	2.21	2.10	2.01	1.94	1.88

Table 6 CRITICAL VALUES OF THE F-DISTRIBUTION
(Description: p. 11)

$\alpha = 0.05$

$f_2 \backslash f_1$	10	12	15	20	24	30	40	60	120
2	19.40	19.41	19.43	19.45	19.45	19.46	19.47	19.48	19.49
3	8.79	8.74	8.70	8.66	8.64	8.62	8.59	8.57	8.55
4	5.96	5.91	5.86	5.80	5.77	5.75	5.72	5.69	5.66
5	4.74	4.68	4.62	4.56	4.53	4.50	4.46	4.43	4.40
6	4.06	4.00	3.94	3.87	3.84	3.81	3.77	3.74	3.70
7	3.64	3.57	3.51	3.44	3.41	3.38	3.34	3.30	3.27
8	3.35	3.28	3.22	3.15	3.12	3.08	3.04	3.01	2.97
9	3.14	3.07	3.01	2.94	2.90	2.86	2.83	2.79	2.75
10	2.98	2.91	2.85	2.77	2.74	2.70	2.66	2.62	2.58
11	2.85	2.79	2.72	2.65	2.61	2.57	2.53	2.49	2.45
12	2.75	2.69	2.62	2.54	2.51	2.47	2.43	2.38	2.34
13	2.67	2.60	2.53	2.46	2.42	2.38	2.34	2.30	2.25
14	2.60	2.53	2.46	2.39	2.35	2.31	2.27	2.22	2.18
15	2.54	2.48	2.40	2.33	2.29	2.25	2.20	2.16	2.11
16	2.49	2.42	2.35	2.28	2.24	2.19	2.15	2.11	2.06
17	2.45	2.38	2.31	2.23	2.19	2.15	2.10	2.06	2.01
18	2.41	2.34	2.27	2.19	2.15	2.11	2.06	2.02	1.97
19	2.38	2.31	2.23	2.16	2.11	2.07	2.03	1.98	1.93
20	2.35	2.28	2.20	2.12	2.08	2.04	1.99	1.95	1.90
21	2.32	2.25	2.18	2.10	2.05	2.01	1.96	1.92	1.87
22	2.30	2.23	2.15	2.07	2.03	1.98	1.94	1.89	1.84
23	2.27	2.20	2.13	2.05	2.01	1.96	1.91	1.86	1.81
24	2.25	2.18	2.11	2.03	1.98	1.94	1.89	1.84	1.79
25	2.24	2.16	2.09	2.01	1.96	1.92	1.87	1.82	1.77
26	2.22	2.15	2.07	1.99	1.95	1.90	1.85	1.80	1.75
27	2.20	2.13	2.06	1.97	1.93	1.88	1.84	1.79	1.73
28	2.19	2.12	2.04	1.96	1.91	1.87	1.82	1.77	1.71
29	2.18	2.10	2.03	1.94	1.90	1.85	1.81	1.75	1.70
30	2.16	2.09	2.01	1.93	1.89	1.84	1.79	1.74	1.68
40	2.08	2.00	1.92	1.84	1.79	1.74	1.69	1.64	1.58
48	2.03	1.96	1.88	1.79	1.75	1.70	1.64	1.59	1.52
60	1.99	1.92	1.84	1.75	1.70	1.65	1.59	1.53	1.47
90	1.94	1.86	1.78	1.69	1.64	1.59	1.53	1.46	1.39
120	1.91	1.83	1.75	1.66	1.61	1.55	1.50	1.43	1.35
∞	1.83	1.75	1.67	1.57	1.52	1.46	1.39	1.32	1.22

Table 6 CRITICAL VALUES OF THE F-DISTRIBUTION
(Description: p. 11)

$\alpha = 0.01$

$f_2 \backslash f_1$	1	2	3	4	5	6	7	8	9
2	98.50	99.00	99.17	99.25	99.30	99.33	99.36	99.37	99.39
3	34.12	30.82	29.46	28.71	28.24	27.91	27.67	27.49	27.35
4	21.20	18.00	16.69	15.98	15.52	15.21	14.98	14.80	14.66
5	16.26	13.27	12.06	11.39	10.97	10.67	10.46	10.29	10.16
6	13.75	10.92	9.78	9.15	8.75	8.47	8.26	8.10	7.98
7	12.25	9.55	8.45	7.85	7.46	7.19	6.99	6.84	6.72
8	11.26	8.65	7.59	7.01	6.63	6.37	6.18	6.03	5.91
9	10.56	8.02	6.99	6.42	6.06	5.80	5.61	5.47	5.35
10	10.04	7.56	6.55	5.99	5.64	5.39	5.20	5.06	4.94
11	9.65	7.21	6.22	5.67	5.32	5.07	4.89	4.74	4.63
12	9.33	6.93	5.95	5.41	5.06	4.82	4.64	4.50	4.39
13	9.07	6.70	5.74	5.21	4.86	4.62	4.44	4.30	4.19
14	8.86	6.51	5.56	5.04	4.69	4.46	4.28	4.14	4.03
15	8.68	6.36	5.42	4.89	4.56	4.32	4.14	4.00	3.89
16	8.53	6.23	5.29	4.77	4.44	4.20	4.03	3.89	3.78
17	8.40	6.11	5.18	4.67	4.34	4.10	3.93	3.79	3.68
18	8.29	6.01	5.09	4.58	4.25	4.01	3.84	3.71	3.60
19	8.18	5.93	5.01	4.50	4.17	3.94	3.77	3.63	3.52
20	8.10	5.85	4.94	4.43	4.10	3.87	3.70	3.56	3.46
21	8.02	5.78	4.87	4.37	4.04	3.81	3.64	3.51	3.40
22	7.95	5.72	4.82	4.31	3.99	3.76	3.59	3.45	3.35
23	7.88	5.66	4.76	4.26	3.94	3.71	3.54	3.41	3.30
24	7.82	5.61	4.72	4.22	3.90	3.67	3.50	3.36	3.26
25	7.77	5.57	4.68	4.18	3.85	3.63	3.46	3.32	3.22
26	7.72	5.53	4.64	4.14	3.82	3.59	3.42	3.29	3.18
27	7.68	5.49	4.60	4.11	3.78	3.56	3.39	3.26	3.15
28	7.64	5.45	4.57	4.07	3.75	3.53	3.36	3.23	3.12
29	7.60	5.42	4.54	4.04	3.73	3.50	3.33	3.20	3.09
30	7.56	5.39	4.51	4.02	3.70	3.47	3.30	3.17	3.07
40	7.31	5.18	4.31	3.83	3.51	3.29	3.12	2.99	2.89
48	7.19	5.08	4.22	3.74	3.43	3.20	3.04	2.91	2.80
60	7.08	4.98	4.13	3.65	3.34	3.12	2.95	2.82	2.72
90	6.93	4.85	4.01	3.53	3.23	3.01	2.84	2.72	2.61
120	6.85	4.79	3.95	3.48	3.17	2.96	2.79	2.66	2.56
∞	6.63	4.61	3.78	3.32	3.02	2.80	2.64	2.51	2.41

Table 6 CRITICAL VALUES OF THE F-DISTRIBUTION
(Description: p. 11)

$\alpha = 0.01$

f_2 \ f_1	10	12	15	20	24	30	40	60	120
2	99.40	99.42	99.43	99.45	99.46	99.47	99.47	99.48	99.49
3	27.23	27.05	26.87	26.69	26.60	26.50	26.41	26.32	26.22
4	14.55	14.37	14.20	14.02	13.93	13.84	13.75	13.65	13.56
5	10.05	9.89	9.72	9.55	9.47	9.38	9.29	9.20	9.11
6	7.87	7.72	7.56	7.40	7.31	7.23	7.14	7.06	6.97
7	6.62	6.47	6.31	6.16	6.07	5.99	5.91	5.82	5.74
8	5.81	5.67	5.52	5.36	5.28	5.20	5.12	5.03	4.95
9	5.26	5.11	4.96	4.81	4.73	4.65	4.57	4.48	4.40
10	4.85	4.71	4.56	4.41	4.33	4.25	4.17	4.08	4.00
11	4.54	4.40	4.25	4.10	4.02	3.94	3.86	3.78	3.69
12	4.30	4.16	4.01	3.86	3.78	3.70	3.62	3.54	3.45
13	4.10	3.96	3.82	3.66	3.59	3.51	3.43	3.34	3.25
14	3.94	3.80	3.66	3.51	3.43	3.35	3.27	3.18	3.09
15	3.80	3.67	3.52	3.37	3.29	3.21	3.13	3.05	2.96
16	3.69	3.55	3.41	3.26	3.18	3.10	3.02	2.93	2.84
17	3.59	3.46	3.31	3.16	3.08	3.00	2.92	2.83	2.75
18	3.51	3.37	3.23	3.08	3.00	2.92	2.84	2.75	2.66
19	3.43	3.30	3.15	3.00	2.92	2.84	2.76	2.67	2.58
20	3.37	3.23	3.09	2.94	2.86	2.78	2.69	2.61	2.52
21	3.31	3.17	3.03	2.88	2.80	2.72	2.64	2.55	2.46
22	3.26	3.12	2.98	2.83	2.75	2.67	2.58	2.50	2.40
23	3.21	3.07	2.93	2.78	2.70	2.62	2.54	2.45	2.35
24	3.17	3.03	2.89	2.74	2.66	2.58	2.49	2.40	2.31
25	3.13	2.99	2.85	2.70	2.62	2.54	2.45	2.36	2.27
26	3.09	2.96	2.81	2.66	2.58	2.50	2.42	2.33	2.23
27	3.06	2.93	2.78	2.63	2.55	2.47	2.38	2.29	2.20
28	3.03	2.90	2.75	2.60	2.52	2.44	2.35	2.26	2.17
29	3.00	2.87	2.73	2.57	2.49	2.41	2.33	2.23	2.14
30	2.98	2.84	2.70	2.55	2.47	2.39	2.30	2.21	2.11
40	2.80	2.66	2.52	2.37	2.29	2.20	2.11	2.02	1.92
48	2.71	2.58	2.44	2.28	2.20	2.12	2.02	1.93	1.82
60	2.63	2.50	2.35	2.20	2.12	2.03	1.94	1.84	1.73
90	2.52	2.39	2.24	2.09	2.00	1.92	1.82	1.72	1.60
120	2.47	2.34	2.19	2.03	1.95	1.86	1.76	1.66	1.53
∞	2.32	2.18	2.04	1.88	1.79	1.70	1.59	1.47	1.32

Table 7 SAMPLE SIZE: ONE-WAY ANALYSIS OF VARIANCE
(Table: pp. 21-23)

Consider k groups with independent samples of size n, normally distributed with the same standard deviation σ. The null hypothesis of equality of the means is tested by an F-test with $k - 1$ and $k(n - 1)$ degrees of freedom at significance level α. If the maximum difference between any two population means is $\Delta\sigma$, the F-test has minimum power when the other $k - 2$ means are midway between the two extremes. Let β be the probability of type II error in this case. Tabulated (for given k, n, α and β) is the Δ that gives this β.

Table 7 SAMPLE SIZE: ONE-WAY ANALYSIS OF VARIANCE
(Description: p. 20)

$\alpha = 0.01$

n \ β	k = 3				k = 4				k = 5			
	.30	.10	.05	.01	.30	.10	.05	.01	.30	.10	.05	.01
4	3.2	3.9	4.3	5.0	3.2	4.0	4.3	5.0	3.3	4.0	4.4	5.0
5	2.6	3.3	3.6	4.1	2.7	3.3	3.6	4.2	2.8	3.4	3.7	4.2
6	2.3	2.8	3.1	3.6	2.4	2.9	3.2	3.7	2.5	3.0	3.3	3.7
7	2.1	2.6	2.8	3.2	2.2	2.6	2.9	3.3	2.2	2.7	3.0	3.4
8	1.9	2.3	2.6	2.9	2.0	2.4	2.6	3.0	2.1	2.5	2.7	3.1
9	1.8	2.2	2.4	2.7	1.9	2.3	2.5	2.8	1.9	2.3	2.5	2.9
10	1.7	2.0	2.2	2.6	1.7	2.1	2.3	2.7	1.8	2.2	2.4	2.7
12	1.5	1.8	2.0	2.3	1.6	1.9	2.1	2.4	1.6	2.0	2.1	2.4
14	1.4	1.7	1.8	2.1	1.4	1.8	1.9	2.2	1.5	1.8	2.0	2.2
16	1.3	1.6	1.7	2.0	1.3	1.6	1.8	2.0	1.4	1.7	1.8	2.1
18	1.2	1.5	1.6	1.8	1.3	1.5	1.7	1.9	1.3	1.6	1.7	2.0
20	1.1	1.4	1.5	1.7	1.2	1.4	1.6	1.8	1.2	1.5	1.6	1.8
30	.90	1.1	1.2	1.4	.95	1.2	1.3	1.4	.99	1.2	1.3	1.5
40	.78	.95	1.0	1.2	.82	1.0	1.1	1.2	.85	1.0	1.1	1.3
50	.69	.85	.92	1.1	.73	.89	.97	1.1	.76	.92	1.0	1.1
80	.54	.67	.73	.84	.57	.70	.76	.87	.60	.73	.79	.90
100	.48	.59	.65	.75	.51	.62	.68	.78	.54	.65	.70	.80
200	.34	.42	.46	.53	.36	.44	.48	.55	.38	.46	.49	.57
300	.28	.34	.37	.43	.29	.36	.39	.45	.31	.37	.40	.46
500	.22	.26	.29	.33	.23	.28	.30	.35	.24	.29	.31	.36

n \ β	k = 6				k = 7				k = 8			
	.30	.10	.05	.01	.30	.10	.05	.01	.30	.10	.05	.01
4	3.4	4.1	4.4	5.1	3.4	4.1	4.5	5.1	3.5	4.2	4.5	5.2
5	2.9	3.5	3.8	4.3	2.9	3.5	3.8	4.4	3.0	3.6	3.9	4.4
6	2.5	3.1	3.3	3.8	2.6	3.1	3.4	3.9	2.6	3.2	3.4	3.9
7	2.3	2.8	3.0	3.4	2.4	2.8	3.1	3.5	2.4	2.9	3.1	3.6
8	2.1	2.6	2.8	3.2	2.2	2.6	2.8	3.2	2.2	2.7	2.9	3.3
9	2.0	2.4	2.6	3.0	2.0	2.4	2.6	3.0	2.1	2.5	2.7	3.1
10	1.9	2.3	2.4	2.8	1.9	2.3	2.5	2.8	2.0	2.3	2.5	2.9
12	1.7	2.0	2.2	2.5	1.7	2.1	2.2	2.6	1.8	2.1	2.3	2.6
14	1.5	1.9	2.0	2.3	1.6	1.9	2.1	2.3	1.6	1.9	2.1	2.4
16	1.4	1.7	1.9	2.1	1.5	1.8	1.9	2.2	1.5	1.8	2.0	2.2
18	1.3	1.6	1.8	2.0	1.4	1.7	1.8	2.0	1.4	1.7	1.8	2.1
20	1.3	1.5	1.7	1.9	1.3	1.6	1.7	1.9	1.3	1.6	1.7	2.0
30	1.0	1.2	1.3	1.5	1.1	1.3	1.4	1.6	1.1	1.3	1.4	1.6
40	.88	1.1	1.2	1.3	.91	1.1	1.2	1.3	.93	1.1	1.2	1.4
50	.79	.95	1.0	1.2	.81	.97	1.1	1.2	.83	1.0	1.1	1.2
80	.62	.75	.81	.92	.64	.77	.83	.94	.65	.78	.85	.96
100	.55	.67	.72	.82	.57	.69	.74	.84	.58	.70	.76	.86
200	.39	.47	.51	.58	.40	.48	.52	.59	.41	.49	.53	.61
300	.32	.38	.42	.47	.33	.39	.43	.48	.34	.40	.43	.49
500	.25	.30	.32	.37	.25	.30	.33	.37	.26	.31	.34	.38

Table 7 SAMPLE SIZE: ONE-WAY ANALYSIS OF VARIANCE
(Description: p. 20)

$\alpha = 0.05$

$n \backslash \beta$	k = 3				k = 4				k = 5			
	.30	.10	.05	.01	.30	.10	.05	.01	.30	.10	.05	.01
4	2.3	3.0	3.3	3.9	2.5	3.1	3.5	4.1	2.6	3.3	3.6	4.2
5	2.0	2.6	2.8	3.4	2.1	2.7	3.0	3.5	2.2	2.8	3.1	3.6
6	1.8	2.3	2.5	3.0	1.9	2.4	2.6	3.1	2.0	2.5	2.7	3.2
7	1.6	2.1	2.3	2.7	1.7	2.2	2.4	2.8	1.8	2.3	2.5	2.9
8	1.5	1.9	2.2	2.5	1.6	2.0	2.2	2.6	1.7	2.1	2.3	2.7
9	1.4	1.8	2.0	2.3	1.5	1.9	2.1	2.4	1.6	2.0	2.2	2.5
10	1.3	1.7	1.9	2.2	1.4	1.8	2.0	2.3	1.5	1.9	2.0	2.4
12	1.2	1.5	1.7	2.0	1.3	1.6	1.8	2.1	1.3	1.7	1.8	2.1
14	1.1	1.4	1.5	1.8	1.2	1.5	1.6	1.9	1.2	1.5	1.7	2.0
16	1.0	1.3	1.4	1.7	1.1	1.4	1.5	1.8	1.1	1.4	1.6	1.8
18	.95	1.2	1.3	1.6	1.0	1.3	1.4	1.7	1.1	1.3	1.5	1.7
20	.90	1.2	1.3	1.5	.96	1.2	1.3	1.6	1.0	1.3	1.4	1.6
30	.73	.93	1.0	1.2	.78	.99	1.1	1.3	.82	1.0	1.1	1.3
40	.63	.80	.89	1.0	.67	.85	.94	1.1	.70	.89	.98	1.1
50	.56	.72	.79	.93	.60	.76	.84	.98	.63	.79	.87	1.0
80	.44	.57	.63	.74	.47	.60	.66	.77	.49	.62	.69	.80
100	.39	.51	.56	.66	.42	.54	.59	.69	.44	.56	.61	.71
200	.28	.36	.39	.46	.30	.38	.42	.49	.31	.39	.43	.50
300	.23	.29	.32	.38	.24	.31	.34	.40	.25	.32	.35	.41
500	.18	.23	.25	.29	.19	.24	.26	.31	.20	.25	.27	.32

$n \backslash \beta$	k = 6				k = 7				k = 8			
	.30	.10	.05	.01	.30	.10	.05	.01	.30	.10	.05	.01
4	2.7	3.3	3.7	4.3	2.7	3.4	3.7	4.3	2.8	3.5	3.8	4.4
5	2.3	2.9	3.2	3.7	2.4	2.9	3.2	3.7	2.4	3.0	3.3	3.8
6	2.0	2.6	2.8	3.3	2.1	2.6	2.9	3.3	2.2	2.7	2.9	3.4
7	1.9	2.3	2.6	3.0	1.9	2.4	2.6	3.1	2.0	2.5	2.7	3.1
8	1.7	2.2	2.4	2.8	1.8	2.2	2.4	2.8	1.8	2.3	2.5	2.9
9	1.6	2.0	2.2	2.6	1.7	2.1	2.3	2.6	1.7	2.1	2.3	2.7
10	1.5	1.9	2.1	2.4	1.6	2.0	2.1	2.5	1.6	2.0	2.2	2.5
12	1.4	1.7	1.9	2.2	1.4	1.8	1.9	2.3	1.5	1.8	2.0	2.3
14	1.3	1.6	1.7	2.0	1.3	1.6	1.8	2.1	1.3	1.7	1.8	2.1
16	1.2	1.5	1.6	1.9	1.2	1.5	1.7	1.9	1.2	1.6	1.7	2.0
18	1.1	1.4	1.5	1.8	1.1	1.4	1.6	1.8	1.2	1.5	1.6	1.9
20	1.0	1.3	1.4	1.7	1.1	1.4	1.5	1.7	1.1	1.4	1.5	1.8
30	.85	1.1	1.2	1.4	.88	1.1	1.2	1.4	.90	1.1	1.2	1.4
40	.73	.92	1.0	1.2	.75	.94	1.0	1.2	.78	.97	1.1	1.2
50	.65	.82	.90	1.0	.67	.84	.92	1.1	.69	.86	.94	1.1
80	.51	.65	.71	.82	.53	.66	.73	.84	.55	.68	.74	.86
100	.46	.58	.63	.73	.47	.59	.65	.75	.49	.61	.66	.77
200	.32	.41	.45	.52	.33	.42	.46	.53	.34	.43	.47	.54
300	.26	.33	.36	.42	.27	.34	.37	.43	.28	.35	.38	.44
500	.20	.26	.28	.33	.21	.26	.29	.34	.22	.27	.30	.34

Table 7 SAMPLE SIZE: ONE-WAY ANALYSIS OF VARIANCE
(Description: p. 20)

$\alpha = 0.10$

	k = 3				k = 4				k = 5			
n \ β	.30	.10	.05	.01	.30	.10	.05	.01	.30	.10	.05	.01
4	2.0	2.6	2.9	3.5	2.1	2.8	3.1	3.7	2.2	2.9	3.2	3.8
5	1.7	2.3	2.5	3.0	1.8	2.4	2.7	3.2	1.9	2.5	2.8	3.3
6	1.5	2.0	2.3	2.7	1.6	2.1	2.4	2.8	1.7	2.2	2.5	2.9
7	1.4	1.8	2.1	2.5	1.5	2.0	2.2	2.6	1.6	2.0	2.3	2.7
8	1.3	1.7	1.9	2.3	1.4	1.8	2.0	2.4	1.5	1.9	2.1	2.5
9	1.2	1.6	1.8	2.1	1.3	1.7	1.9	2.2	1.4	1.8	2.0	2.3
10	1.1	1.5	1.7	2.0	1.2	1.6	1.8	2.1	1.3	1.7	1.9	2.2
12	1.0	1.4	1.5	1.8	1.1	1.5	1.6	1.9	1.2	1.5	1.7	2.0
14	.95	1.3	1.4	1.7	1.0	1.3	1.5	1.8	1.1	1.4	1.5	1.8
16	.89	1.2	1.3	1.6	.95	1.2	1.4	1.6	1.0	1.3	1.4	1.7
18	.83	1.1	1.2	1.5	.89	1.2	1.3	1.5	.94	1.2	1.4	1.6
20	.79	1.0	1.2	1.4	.85	1.1	1.2	1.5	.89	1.2	1.3	1.5
30	.64	.85	.94	1.1	.69	.90	1.0	1.2	.72	.94	1.0	1.2
40	.55	.73	.81	.97	.59	.78	.86	1.0	.62	.81	.90	1.1
50	.49	.65	.73	.87	.53	.69	.77	.91	.56	.72	.80	.95
80	.39	.51	.57	.68	.42	.55	.61	.72	.44	.57	.63	.75
100	.35	.46	.51	.61	.37	.49	.54	.64	.39	.51	.56	.67
200	.24	.32	.36	.43	.26	.34	.38	.45	.28	.36	.40	.47
300	.20	.26	.30	.35	.21	.28	.31	.37	.23	.29	.33	.38
500	.15	.20	.23	.27	.17	.22	.24	.29	.17	.23	.25	.30

	k = 6				k = 7				k = 8			
n \ β	.30	.10	.05	.01	.30	.10	.05	.01	.30	.10	.05	.01
4	2.3	3.0	3.3	3.9	2.4	3.1	3.4	4.0	2.4	3.1	3.5	4.1
5	2.0	2.6	2.9	3.4	2.1	2.7	2.9	3.4	2.1	2.7	3.0	3.5
6	1.8	2.3	2.6	3.0	1.8	2.4	2.6	3.1	1.9	2.4	2.7	3.2
7	1.6	2.1	2.3	2.8	1.7	2.2	2.4	2.8	1.7	2.2	2.5	2.9
8	1.5	2.0	2.2	2.6	1.6	2.0	2.2	2.6	1.6	2.1	2.3	2.7
9	1.4	1.8	2.0	2.4	1.5	1.9	2.1	2.5	1.5	1.9	2.1	2.5
10	1.3	1.7	1.9	2.3	1.4	1.8	2.0	2.3	1.4	1.8	2.0	2.4
12	1.2	1.6	1.7	2.0	1.3	1.6	1.8	2.1	1.3	1.7	1.8	2.1
14	1.1	1.4	1.6	1.9	1.2	1.5	1.6	1.9	1.2	1.5	1.7	2.0
16	1.0	1.3	1.5	1.8	1.1	1.4	1.5	1.8	1.1	1.4	1.6	1.8
18	.98	1.3	1.4	1.7	1.0	1.3	1.4	1.7	1.0	1.3	1.5	1.7
20	.93	1.2	1.3	1.6	.96	1.2	1.4	1.6	.99	1.3	1.4	1.6
30	.75	.97	1.1	1.3	.78	1.0	1.1	1.3	.80	1.0	1.1	1.3
40	.65	.84	.93	1.1	.67	.86	.95	1.1	.69	.89	.98	1.1
50	.58	.75	.83	.98	.60	.77	.85	1.0	.62	.79	.87	1.0
80	.46	.59	.65	.77	.47	.61	.67	.79	.49	.62	.69	.81
100	.41	.53	.58	.69	.42	.54	.60	.70	.43	.56	.61	.72
200	.29	.37	.41	.49	.30	.38	.42	.50	.31	.39	.43	.51
300	.24	.30	.34	.40	.24	.31	.35	.41	.25	.32	.35	.41
500	.18	.24	.26	.31	.19	.24	.27	.31	.19	.25	.27	.32

Table 8 ONE-SIDED TOLERANCE LIMITS (Table: p. 25)

For a sample of size n of a normal random variable X, consider the sample mean $\bar{x}$ and the sample standard deviation s

$$\bar{x} = \sum_{i=1}^{n} \frac{X_i}{n}, \quad s = \sqrt{\frac{\sum_{i=1}^{n}(X_i - \bar{x})^2}{n - 1}}$$

The table entry gives factors k such that with confidence γ the interval $(-\infty, \bar{x} + ks)$ contains at least the proportion P of the normal distribution being sampled. The same k may be used for tolerance intervals of the form $(\bar{x} - ks, +\infty)$.

Table 8 ONE-SIDED TOLERANCE LIMITS (Description: p. 24)

$\gamma = .90$

n	P .900	.950	.990	.999
3	4.26	5.31	7.34	9.65
4	3.19	3.96	5.44	7.13
5	2.74	3.40	4.67	6.11
6	2.49	3.09	4.24	5.56
7	2.33	2.89	3.97	5.20
8	2.22	2.75	3.78	4.96
9	2.13	2.65	3.64	4.77
10	2.07	2.57	3.53	4.63
11	2.01	2.50	3.44	4.51
12	1.97	2.45	3.37	4.42
13	1.93	2.40	3.31	4.34
14	1.90	2.36	3.26	4.27
15	1.87	2.33	3.21	4.22
16	1.84	2.30	3.17	4.16
17	1.82	2.27	3.14	4.12
18	1.80	2.25	3.11	4.08
19	1.78	2.23	3.08	4.04
20	1.77	2.21	3.05	4.01
21	1.75	2.19	3.03	3.98
22	1.74	2.17	3.01	3.95
23	1.72	2.16	2.99	3.93
24	1.71	2.15	2.97	3.90
25	1.70	2.13	2.95	3.88
30	1.66	2.08	2.88	3.79
35	1.62	2.04	2.83	3.73
40	1.60	2.01	2.79	3.68
45	1.58	1.99	2.76	3.64
50	1.56	1.97	2.73	3.60
60	1.53	1.93	2.69	3.55
70	1.51	1.91	2.66	3.51
80	1.49	1.89	2.64	3.48
90	1.48	1.87	2.62	3.46
100	1.47	1.86	2.60	3.44
120	1.45	1.84	2.57	3.40
145	1.44	1.82	2.55	3.37
300	1.39	1.76	2.48	3.28
500	1.36	1.74	2.44	3.24
∞	1.28	1.64	2.33	3.09

$\gamma = .95$

n	P .900	.950	.990	.999
3	6.16	7.66	10.55	13.86
4	4.16	5.14	7.04	9.21
5	3.41	4.20	5.74	7.50
6	3.01	3.71	5.06	6.61
7	2.76	3.40	4.64	6.06
8	2.58	3.19	4.35	5.69
9	2.45	3.03	4.14	5.41
10	2.35	2.91	3.98	5.20
11	2.28	2.81	3.85	5.04
12	2.21	2.74	3.75	4.90
13	2.16	2.67	3.66	4.79
14	2.11	2.61	3.58	4.69
15	2.07	2.57	3.52	4.61
16	2.03	2.52	3.46	4.54
17	2.00	2.49	3.41	4.47
18	1.97	2.45	3.37	4.41
19	1.95	2.42	3.33	4.36
20	1.93	2.40	3.30	4.32
21	1.91	2.37	3.26	4.28
22	1.89	2.35	3.23	4.24
23	1.87	2.33	3.21	4.20
24	1.85	2.31	3.18	4.17
25	1.84	2.29	3.16	4.14
30	1.78	2.22	3.06	4.02
35	1.73	2.17	2.99	3.93
40	1.70	2.13	2.94	3.87
45	1.67	2.09	2.90	3.81
50	1.65	2.06	2.86	3.77
60	1.61	2.02	2.81	3.70
70	1.58	1.99	2.77	3.64
80	1.56	1.96	2.73	3.60
90	1.54	1.94	2.71	3.57
100	1.53	1.93	2.68	3.54
120	1.50	1.90	2.65	3.50
145	1.48	1.87	2.62	3.45
300	1.42	1.80	2.52	3.34
500	1.39	1.76	2.48	3.28
∞	1.28	1.64	2.33	3.09

Table 9 TWO-SIDED TOLERANCE LIMITS (Table: p. 27)

The sample mean $\bar{x}$ and sample standard deviation s are calculated for a normal sample of size n (as for Table 8). In this table, values k are given such that with confidence γ the interval $(\bar{x} - ks, \bar{x} + ks)$ contains at least the proportion P of the normal distribution sampled.

Table 9 TWO-SIDED TOLERANCE LIMITS (Description: p. 26)

$\gamma = .90$

n	P = .900	.950	.990	.999
3	5.85	6.92	8.97	11.31
4	4.17	4.94	6.44	8.15
5	3.49	4.15	5.42	6.88
6	3.13	3.72	4.87	6.19
7	2.90	3.45	4.52	5.75
8	2.74	3.26	4.28	5.45
9	2.63	3.13	4.10	5.22
10	2.54	3.02	3.96	5.05
11	2.46	2.93	3.85	4.91
12	2.40	2.86	3.76	4.79
13	2.36	2.81	3.68	4.70
14	2.31	2.76	3.62	4.62
15	2.28	2.71	3.56	4.55
16	2.25	2.68	3.51	4.48
17	2.22	2.64	3.47	4.43
18	2.19	2.61	3.43	4.38
19	2.17	2.59	3.40	4.34
20	2.15	2.56	3.37	4.30
21	2.13	2.54	3.34	4.26
22	2.12	2.52	3.32	4.23
23	2.10	2.51	3.29	4.20
24	2.09	2.49	3.27	4.18
25	2.08	2.47	3.25	4.15
30	2.03	2.41	3.17	4.05
35	1.99	2.37	3.11	3.97
40	1.96	2.33	3.07	3.92
45	1.94	2.31	3.03	3.87
50	1.92	2.28	3.00	3.83
60	1.89	2.25	2.95	3.77
70	1.87	2.22	2.92	3.73
80	1.85	2.20	2.89	3.70
90	1.83	2.19	2.87	3.67
100	1.82	2.17	2.85	3.65
120	1.80	2.15	2.83	3.61
145	1.79	2.13	2.80	3.58
300	1.74	2.07	2.72	3.48
500	1.72	2.05	2.69	3.43
∞	1.64	1.96	2.58	3.29

$\gamma = .95$

n	P = .900	.950	.990	.999
3	8.38	9.92	12.86	16.21
4	5.37	6.37	8.30	10.50
5	4.27	5.08	6.63	8.42
6	3.71	4.41	5.77	7.34
7	3.37	4.01	5.25	6.68
8	3.14	3.73	4.89	6.23
9	2.97	3.53	4.63	5.90
10	2.84	3.38	4.43	5.65
11	2.74	3.26	4.28	5.45
12	2.65	3.16	4.15	5.29
13	2.59	3.08	4.04	5.16
14	2.53	3.01	3.95	5.05
15	2.48	2.95	3.88	4.95
16	2.44	2.90	3.81	4.86
17	2.40	2.86	3.75	4.79
18	2.37	2.82	3.70	4.73
19	2.34	2.78	3.66	4.67
20	2.31	2.75	3.61	4.61
21	2.29	2.72	3.58	4.57
22	2.26	2.70	3.54	4.52
23	2.24	2.67	3.51	4.48
24	2.23	2.65	3.48	4.45
25	2.21	2.63	3.46	4.41
30	2.14	2.55	3.35	4.28
35	2.09	2.49	3.27	4.18
40	2.05	2.44	3.21	4.10
45	2.02	2.41	3.16	4.04
50	2.00	2.38	3.13	3.99
60	1.96	2.33	3.07	3.92
70	1.93	2.30	3.02	3.86
80	1.91	2.27	2.99	3.81
90	1.89	2.25	2.96	3.78
100	1.87	2.23	2.93	3.75
120	1.85	2.20	2.90	3.70
145	1.83	2.18	2.86	3.66
300	1.77	2.11	2.77	3.53
500	1.74	2.07	2.72	3.48
∞	1.64	1.96	2.58	3.29

Table 10 FACTORS FOR TOLERANCE LIMITS CONTROLLING BOTH TAILS OF A NORMAL DISTRIBUTION (Table: p. 29)

The sample mean $\bar{x}$ and sample standard deviation s are calculated as for Table 8 (see p. 24). For given n and γ, let two probabilities P_1 and P_2 be prescribed, and let k_1 and k_2 be the corresponding values in Table 10. Then with confidence γ the interval $(-\infty, \bar{x} - k_1 s)$ contains not more than the proportion P_1, and the interval $(\bar{x} + k_2 s, \infty)$ not more than the proportion P_2, of the normal distribution sampled. Example: $\gamma = 0.95$, $n = 50$, $P_1 = 0.01$, $P_2 = 0.05$; then $k_1 = 3.26$ and $k_2 = 2.52$.

Table 10 FACTORS FOR TOLERANCE LIMITS CONTROLLING BOTH TAILS OF A NORMAL DISTRIBUTION (Description: p. 28)

$\gamma = .90$

	P			
n	0.10	0.05	0.01	0.001
3	6.55	7.52	9.40	11.59
4	4.70	5.40	6.79	8.40
5	3.95	4.55	5.73	7.11
6	3.54	4.08	5.16	6.41
7	3.27	3.78	4.79	5.96
8	3.09	3.57	4.53	5.65
9	2.95	3.42	4.34	5.42
10	2.84	3.30	4.19	5.23
11	2.75	3.20	4.07	5.09
12	2.68	3.12	3.97	4.97
13	2.62	3.05	3.89	4.87
14	2.57	2.99	3.82	4.79
15	2.53	2.94	3.76	4.71
16	2.49	2.90	3.71	4.65
17	2.45	2.86	3.66	4.59
18	2.42	2.82	3.62	4.54
19	2.39	2.79	3.58	4.49
20	2.37	2.76	3.54	4.45
21	2.34	2.74	3.51	4.41
22	2.32	2.72	3.48	4.38
23	2.30	2.69	3.46	4.35
24	2.29	2.67	3.43	4.32
25	2.27	2.65	3.41	4.29
30	2.20	2.58	3.32	4.18
35	2.15	2.52	3.25	4.10
40	2.11	2.48	3.20	4.03
45	2.08	2.44	3.15	3.98
50	2.05	2.41	3.12	3.94
60	2.01	2.37	3.06	3.87
70	1.98	2.33	3.02	3.82
80	1.96	2.31	2.99	3.78
90	1.94	2.28	2.96	3.75
100	1.92	2.27	2.94	3.73
120	1.89	2.24	2.91	3.68
145	1.87	2.21	2.87	3.64
300	1.80	2.13	2.78	3.53
500	1.76	2.09	2.73	3.47
∞	1.64	1.96	2.58	3.29

$\gamma = .95$

	P			
n	0.10	0.05	0.01	0.001
3	9.41	10.79	13.49	16.63
4	6.07	6.98	8.76	10.83
5	4.85	5.58	7.03	8.71
6	4.21	4.86	6.13	7.61
7	3.82	4.41	5.57	6.93
8	3.55	4.10	5.19	6.47
9	3.35	3.88	4.92	6.13
10	3.20	3.70	4.70	5.87
11	3.08	3.57	4.54	5.66
12	2.98	3.46	4.40	5.50
13	2.90	3.36	4.28	5.36
14	2.83	3.28	4.19	5.24
15	2.77	3.22	4.10	5.14
16	2.71	3.16	4.03	5.05
17	2.67	3.10	3.97	4.97
18	2.62	3.06	3.91	4.90
19	2.59	3.02	3.86	4.84
20	2.55	2.98	3.81	4.78
21	2.52	2.94	3.77	4.73
22	2.50	2.91	3.73	4.69
23	2.47	2.88	3.70	4.64
24	2.45	2.86	3.66	4.60
25	2.43	2.83	3.63	4.57
30	2.34	2.73	3.51	4.42
35	2.27	2.66	3.42	4.31
40	2.22	2.60	3.36	4.23
45	2.18	2.56	3.30	4.16
50	2.15	2.52	3.26	4.11
60	2.10	2.46	3.18	4.02
70	2.06	2.42	3.13	3.96
80	2.03	2.39	3.09	3.91
90	2.00	2.36	3.06	3.87
100	1.98	2.33	3.03	3.83
120	1.95	2.30	2.98	3.78
145	1.92	2.26	2.94	3.73
300	1.83	2.16	2.82	3.59
500	1.79	2.12	2.76	3.52
∞	1.64	1.96	2.58	3.29

Table 11 CRITICAL VALUES OF THE SAMPLE RANGE FROM A NORMAL DISTRIBUTION (Table: p. 31)

Let $X_1, \ldots, X_n$ be a random sample from $N(\mu, \sigma^2)$. Then the range of this sample is defined by

$$R = \max(X_1, \ldots, X_n) - \min(X_1, \ldots, X_n)$$

The table contains values w such that, for given p, one has

$$P\left(\frac{R}{\sigma} \leq w\right) = p$$

Table 11 CRITICAL VALUES OF THE SAMPLE RANGE FROM A NORMAL DISTRIBUTION (Description: p. 30)

	cumulative probability, p									
n	.005	.01	.025	.05	.10	.90	.95	.975	.99	.995
2	0.01	0.02	0.04	0.09	0.18	2.33	2.77	3.17	3.64	3.97
3	0.13	0.19	0.30	0.43	0.62	2.90	3.31	3.68	4.12	4.42
4	0.34	0.43	0.59	0.76	0.98	3.24	3.63	3.98	4.40	4.69
5	0.55	0.67	0.85	1.03	1.26	3.48	3.86	4.20	4.60	4.89
6	0.75	0.87	1.07	1.25	1.49	3.66	4.03	4.36	4.78	5.03
7	0.92	1.05	1.25	1.44	1.68	3.81	4.17	4.49	4.88	5.15
8	1.08	1.20	1.41	1.60	1.84	3.93	4.29	4.60	4.99	5.25
9	1.21	1.34	1.55	1.74	1.97	4.04	4.39	4.70	5.08	5.34
10	1.33	1.47	1.67	1.86	2.09	4.13	4.47	4.78	5.16	5.42
11	1.45	1.58	1.78	1.97	2.20	4.21	4.55	4.86	5.23	5.49
12	1.55	1.68	1.88	2.07	2.30	4.28	4.62	4.92	5.29	5.55
13	1.64	1.77	1.98	2.16	2.39	4.35	4.68	4.99	5.35	5.60
14	1.72	1.86	2.06	2.24	2.47	4.41	4.74	5.04	5.40	5.65
15	1.80	1.93	2.14	2.32	2.54	4.47	4.80	5.09	5.45	5.70
16	1.88	2.01	2.21	2.39	2.61	4.52	4.85	5.14	5.49	5.74
17	1.94	2.07	2.27	2.45	2.67	4.57	4.89	5.18	5.54	5.78
18	2.01	2.14	2.34	2.52	2.73	4.61	4.93	5.22	5.57	5.82
19	2.07	2.20	2.39	2.57	2.79	4.65	4.97	5.26	5.61	5.86
20	2.13	2.25	2.45	2.63	2.84	4.69	5.01	5.30	5.65	5.89
22	2.23	2.36	2.55	2.72	2.94	4.77	5.08	5.37	5.71	5.95
24	2.32	2.45	2.64	2.81	3.02	4.83	5.14	5.43	5.77	6.01
26	2.41	2.53	2.72	2.89	3.10	4.89	5.20	5.48	5.82	6.06
28	2.49	2.61	2.80	2.97	3.17	4.95	5.25	5.53	5.87	6.10
30	2.56	2.68	2.87	3.03	3.24	5.00	5.30	5.58	5.91	6.15
32	2.63	2.75	2.93	3.10	3.30	5.04	5.35	5.62	5.95	6.19
34	2.69	2.81	2.99	3.16	3.35	5.09	5.39	5.66	5.99	6.22
36	2.75	2.86	3.05	3.21	3.41	5.13	5.43	5.70	6.03	6.26
38	2.80	2.92	3.10	3.26	3.46	5.17	5.46	5.73	6.06	6.29
40	2.85	2.97	3.15	3.31	3.50	5.20	5.50	5.77	6.09	6.32
50	3.07	3.18	3.36	3.51	3.70	5.36	5.65	5.91	6.23	6.45
60	3.24	3.35	3.52	3.68	3.86	5.48	5.76	6.02	6.34	6.56
70	3.39	3.49	3.66	3.81	3.99	5.58	5.86	6.12	6.43	6.65
80	3.51	3.61	3.78	3.92	4.10	5.67	5.95	6.20	6.51	6.73
90	3.61	3.72	3.88	4.02	4.20	5.75	6.02	6.27	6.58	6.79
100	3.70	3.81	3.96	4.11	4.28	5.81	6.08	6.33	6.64	6.85

Table 12 MEAN, VARIANCE, AND STANDARD DEVIATION OF THE RANGE, CONTROL CHART FACTORS (Table: p. 33)

Let $X_1, \ldots, X_n$ be a sample from a normal variable with variance 1, R as for Table 11, and

$$\hat{\sigma} = \sqrt{\frac{\sum_{i=1}^{n} (X_i - \bar{x})^2}{n}}$$

The seven columns of this table contain, respectively, n, E(R), σ^2(R), σ(R), 1/E(R), E($\hat{\sigma}$) and 1/E($\hat{\sigma}$). Uses of range tests and construction of quality control charts are mentioned in the references and in the Table 17 description.

Table 12 MEAN, VARIANCE, AND STANDARD DEVIATION OF THE RANGE, CONTROL CHART FACTORS (Description: p. 32)

n	$E(R)$	$\sigma^2(R)$	$\sigma(R)$	$1/E(R)$	$E(\hat{\sigma})$	$1/E(\hat{\sigma})$
2	1.128	0.727	0.853	0.886	0.564	1.772
3	1.693	0.789	0.888	0.591	0.724	1.382
4	2.059	0.774	0.880	0.486	0.798	1.253
5	2.326	0.747	0.864	0.430	0.841	1.189
6	2.534	0.719	0.848	0.395	0.869	1.151
7	2.704	0.694	0.833	0.370	0.888	1.126
8	2.847	0.672	0.820	0.351	0.903	1.108
9	2.970	0.653	0.808	0.337	0.914	1.094
10	3.078	0.635	0.797	0.325	0.923	1.084
11	3.173	0.620	0.787	0.315	0.930	1.075
12	3.258	0.606	0.778	0.307	0.936	1.068
13	3.336	0.594	0.770	0.300	0.941	1.063
14	3.407	0.582	0.763	0.294	0.945	1.058
15	3.472	0.572	0.756	0.288	0.949	1.054
16	3.532	0.562	0.750	0.283	0.952	1.050
17	3.588	0.554	0.744	0.279	0.955	1.047
18	3.640	0.546	0.739	0.275	0.958	1.044
19	3.689	0.538	0.733	0.271	0.960	1.042
20	3.735	0.531	0.729	0.268	0.962	1.040
21	3.778	0.524	0.724	0.265	0.964	1.038
22	3.819	0.518	0.720	0.262	0.965	1.036
23	3.858	0.512	0.716	0.259	0.967	1.034
24	3.895	0.507	0.712	0.257	0.968	1.033
25	3.931	0.502	0.708	0.254	0.970	1.031

Table 10 CRITICAL VALUES FOR TESTING THE DEVIATION OF THE SAMPLE MEAN FROM A PREASSIGNED VALUE USING THE SAMPLE RANGE (Table: p. 35)

Let $\bar{x}$ be the sample mean and R the sample range (as defined for Table 11, p. 30) of a normal sample. Critical values are given for the use of the statistic $(\bar{x} - \mu_0)/R$ to test that the true mean is μ_0. Tabulated are values C (for given sample size n and γ) such that if $E(\bar{x}) = \mu_0$, then $P\{(\bar{x} - \mu_0)/R \leq C\} = \gamma$. The statistic is symmetric so that a two-sided test at significance level α results if γ satisfies $\alpha = 2(1 - \gamma)$ and the rejection region is $|\bar{x} - \mu_0|/R \geq C$ where C is the tabulated critical value.

Table 13 CRITICAL VALUES FOR TESTING THE DEVIATION OF THE SAMPLE MEAN FROM A PREASSIGNED VALUE USING THE SAMPLE RANGE (Description: p. 34)

	cumulative probability, γ					
n	0.95	0.975	0.99	0.995	0.999	0.9995
2	3.157	6.353	15.910	31.828	159.154	318.310
3	0.885	1.304	2.111	3.008	6.768	9.578
4	0.529	0.717	1.023	1.317	2.304	2.916
5	0.388	0.507	0.685	0.842	1.314	1.578
6	0.312	0.399	0.523	0.628	0.921	1.074
7	0.263	0.333	0.429	0.507	0.715	0.819
8	0.230	0.288	0.366	0.429	0.589	0.667
9	0.205	0.255	0.322	0.374	0.504	0.566
10	0.186	0.230	0.288	0.333	0.444	0.495
11	0.170	0.210	0.262	0.302	0.397	0.441
12	0.158	0.194	0.241	0.277	0.361	0.399
13	0.147	0.181	0.224	0.256	0.332	0.366
14	0.138	0.170	0.209	0.239	0.308	0.339
15	0.131	0.160	0.197	0.224	0.288	0.316
16	0.124	0.151	0.186	0.212	0.271	0.296
17	0.118	0.144	0.177	0.201	0.256	0.279
18	0.113	0.137	0.168	0.191	0.243	0.265
19	0.108	0.132	0.161	0.182	0.231	0.252
20	0.104	0.126	0.154	0.175	0.221	0.240

Table 14 CRITICAL VALUES FOR TESTING THE SIGNIFICANCE OF THE DIFFERENCE BETWEEN MEANS OF TWO SAMPLES OF EQUAL SIZE USING THE SAMPLE RANGES (Table: p. 37)

Let $\bar{x}_1$, R_1 and $\bar{x}_2$, R_2 be the sample mean and range, respectively, of two samples of size n from two normal populations with equal variance. The statistic $G = 2|\bar{x}_1 - \bar{x}_2|/(R_1 + R_2)$ is used to test equality of population means. Tabulated are critical values, g, such that if the two population means are equal then $P(G \le g) = \gamma$ for the specified sample size n. That is, the $\alpha = 1 - \gamma$ critical region is $G \ge g$.

Table 15 CRITICAL VALUES OF THE STUDENTIZED RANGE FROM A NORMAL DISTRIBUTION (Table: pp. 38-39)

Let R be the sample range from a normal sample of size n with variance σ^2. Let s be such that $\nu s^2/\sigma^2$ is a χ^2 variable with ν degrees of freedom and s is independent of R. For specified ν, n, and γ (the cumulative probability), critical values, c, are tabulated such that

$$P\left(\frac{R}{s} \le c\right) = \gamma$$

Table 14 CRITICAL VALUES FOR TESTING THE SIGNIFICANCE OF THE DIFFERENCE BETWEEN MEANS OF TWO SAMPLES OF EQUAL SIZE USING THE SAMPLE RANGE (Description: p. 36)

	cumulative probability, γ					
n	0.95	0.975	0.99	0.995	0.999	0.9995
2	2.322	3.427	5.553	7.916	17.813	25.212
3	0.974	1.272	1.722	2.122	3.363	4.123
4	0.644	0.813	1.047	1.237	1.741	1.994
5	0.493	0.613	0.772	0.896	1.205	1.351
6	0.405	0.499	0.621	0.714	0.937	1.038
7	0.347	0.426	0.525	0.600	0.775	0.853
8	0.306	0.373	0.458	0.521	0.666	0.730
9	0.275	0.334	0.409	0.463	0.588	0.642
10	0.250	0.304	0.371	0.419	0.529	0.575
11	0.231	0.280	0.340	0.384	0.482	0.523
12	0.214	0.260	0.315	0.355	0.444	0.482
13	0.201	0.243	0.294	0.331	0.413	0.447
14	0.189	0.228	0.276	0.311	0.386	0.418
15	0.179	0.216	0.261	0.293	0.364	0.393
16	0.170	0.205	0.248	0.278	0.344	0.372
17	0.162	0.195	0.236	0.264	0.327	0.353
18	0.155	0.187	0.225	0.252	0.312	0.336
19	0.149	0.179	0.216	0.242	0.298	0.321
20	0.143	0.172	0.207	0.232	0.286	0.308

Table 15 CRITICAL VALUES OF THE STUDENTIZED RANGE FROM A NORMAL DISTRIBUTION (Description: p. 36)

cumulative probability, γ = 0.90

ν	n=3	n=4	n=5	n=6	n=7	n=8	n=9	n=10
2	5.73	6.77	7.54	8.14	8.63	9.05	9.41	9.72
4	3.98	4.59	5.03	5.39	5.68	5.93	6.14	6.33
6	3.56	4.07	4.44	4.73	4.97	5.17	5.34	5.50
8	3.37	3.83	4.17	4.43	4.65	4.83	4.99	5.13
10	3.27	3.70	4.02	4.26	4.47	4.64	4.78	4.91
12	3.20	3.62	3.92	4.16	4.35	4.51	4.65	4.78
14	3.16	3.56	3.85	4.08	4.27	4.42	4.56	4.68
16	3.12	3.52	3.80	4.03	4.21	4.36	4.49	4.61
18	3.10	3.49	3.77	3.98	4.16	4.31	4.44	4.55
20	3.08	3.46	3.74	3.95	4.12	4.27	4.40	4.51
24	3.05	3.42	3.69	3.90	4.07	4.21	4.34	4.44
30	3.02	3.39	3.65	3.85	4.02	4.16	4.28	4.38
40	2.99	3.35	3.60	3.80	3.96	4.10	4.21	4.32
60	2.97	3.31	3.56	3.75	3.91	4.04	4.16	4.25
120	2.93	3.28	3.52	3.71	3.86	3.99	4.10	4.19
∞	2.90	3.24	3.48	3.66	3.81	3.93	4.04	4.13

cumulative probability, γ = 0.95

ν	n=3	n=4	n=5	n=6	n=7	n=8	n=9	n=10
2	8.33	9.80	10.88	11.73	12.43	13.03	13.54	13.99
4	5.04	5.76	6.29	6.71	7.05	7.35	7.60	7.83
6	4.34	4.90	5.30	5.63	5.90	6.12	6.32	6.49
8	4.04	4.53	4.89	5.17	5.40	5.60	5.77	5.92
10	3.88	4.33	4.65	4.91	5.12	5.30	5.46	5.60
12	3.77	4.20	4.51	4.75	4.95	5.12	5.27	5.39
14	3.70	4.11	4.41	4.64	4.83	4.99	5.13	5.25
16	3.65	4.05	4.33	4.56	4.74	4.90	5.03	5.15
18	3.61	4.00	4.28	4.49	4.67	4.82	4.96	5.07
20	3.58	3.96	4.23	4.45	4.62	4.77	4.90	5.01
24	3.53	3.90	4.17	4.37	4.54	4.68	4.81	4.92
30	3.49	3.85	4.10	4.30	4.46	4.60	4.72	4.82
40	3.44	3.79	4.04	4.23	4.39	4.52	4.63	4.73
60	3.40	3.74	3.98	4.16	4.31	4.44	4.55	4.65
120	3.36	3.68	3.92	4.10	4.24	4.36	4.47	4.56
∞	3.31	3.63	3.86	4.03	4.17	4.29	4.39	4.47

Table 15 CRITICAL VALUES OF THE STUDENTIZED RANGE FROM A NORMAL DISTRIBUTION (Description: p. 36)

cumulative probability, γ = 0.975

ν	n=3	n=4	n=5	n=6	n=7	n=8	n=9	n=10
2	11.94	14.01	15.54	16.75	17.74	18.58	19.31	19.95
4	6.24	7.09	7.72	8.21	8.62	8.98	9.28	9.55
6	5.16	5.77	6.23	6.59	6.88	7.14	7.36	7.55
8	4.71	5.23	5.62	5.92	6.17	6.38	6.57	6.73
10	4.47	4.94	5.29	5.56	5.78	5.97	6.14	6.28
12	4.32	4.76	5.00	5.33	5.54	5.72	5.87	6.00
14	4.22	4.64	4.94	5.18	5.37	5.54	5.68	5.81
16	4.15	4.55	4.84	5.07	5.25	5.41	5.55	5.67
18	4.09	4.48	4.76	4.98	5.16	5.31	5.45	5.57
20	4.05	4.43	4.70	4.91	5.09	5.24	5.37	5.48
24	3.98	4.35	4.61	4.82	4.98	5.13	5.25	5.36
30	3.92	4.27	4.52	4.72	4.88	5.02	5.13	5.24
40	3.86	4.20	4.44	4.63	4.78	4.91	5.02	5.12
60	3.80	4.12	4.36	4.54	4.68	4.81	4.91	5.01
120	3.74	4.05	4.28	4.45	4.59	4.70	4.81	4.89
∞	3.68	3.98	4.20	4.36	4.49	4.60	4.70	4.78

cumulative probability, γ = 0.99

ν	n=3	n=4	n=5	n=6	n=7	n=8	n=9	n=10
2	19.02	22.29	24.72	26.63	28.20	29.53	30.68	31.69
4	8.12	9.17	9.96	10.58	11.10	11.54	11.93	12.26
6	6.33	7.03	7.56	7.97	8.32	8.61	8.87	9.10
8	5.64	6.20	6.62	6.96	7.24	7.47	7.68	7.86
10	5.27	5.77	6.14	6.43	6.67	6.87	7.05	7.21
12	5.05	5.50	5.84	6.10	6.32	6.51	6.67	6.81
14	4.89	5.32	5.63	5.88	6.08	6.26	6.41	6.54
16	4.79	5.19	5.49	5.72	5.92	6.08	6.22	6.35
18	4.70	5.09	5.38	5.60	5.79	5.94	6.08	6.20
20	4.64	5.02	5.29	5.51	5.69	5.84	5.97	6.09
24	4.55	4.91	5.17	5.37	5.54	5.69	5.81	5.92
30	4.45	4.80	5.05	5.24	5.40	5.54	5.65	5.76
40	4.37	4.70	4.93	5.11	5.26	5.39	5.50	5.60
60	4.28	4.59	4.82	4.99	5.13	5.25	5.36	5.45
120	4.20	4.50	4.71	4.87	5.01	5.12	5.21	5.30
∞	4.12	4.40	4.60	4.76	4.88	4.99	5.08	5.16

Table 16 CHI-APPROXIMATION TO THE MEAN RANGE
(Table: p. 41)

Let there be k distinct normal samples, each of size n, from distributions with the same standard deviation σ. Let $R_1, \ldots, R_k$ be the sample ranges (as defined for Table 11, p. 30) of the k samples and define $R = \sum_{i=1}^{k} R_i/k$ to be the mean range. Then there exist a scale factor c and number ν (which are functions of k and n) such that R is approximately distributed as $c\chi\sigma/\sqrt{\nu}$ where χ^2 has a chi square distribution with ν degrees of freedom. This table gives the values of c and ν as functions of n and k.

This table, in conjunction with Table 15 on the Studentized range, may be used for short-cut tests in the analysis of variance and two-sample t-test situations.

1. In the one-way analysis of variance if $\bar{X}_{max}$ and $\bar{X}_{min}$ are the largest and smallest sample means in the k samples, then

$$(\bar{X}_{max} - \bar{X}_{min})c\sqrt{n}/R$$

may be used to test that all means are the same by using Table 15 with a sample size of k and ν degrees of freedom (interpolation needed for the critical value).

2. For two samples of size n use as the test statistic

$$|\bar{X}_1 - \bar{X}_2|c\sqrt{n}/R$$

Table 16 CHI-APPROXIMATION TO THE MEAN RANGE
(Description: p. 40)

k	ν	c	ν	c	ν	c
	n=2		n=3		n=4	
1	1.000	1.414	1.985	1.912	2.929	2.239
2	1.920	1.279	3.834	1.805	5.694	2.151
3	2.817	1.231	5.663	1.769	8.441	2.120
4	3.706	1.206	7.485	1.750	11.185	2.105
5	4.591	1.191	9.305	1.739	13.926	2.096
6	5.473	1.181	11.123	1.731	16.666	2.090
7	6.353	1.173	12.941	1.726	19.405	2.085
8	7.232	1.168	14.757	1.721	22.144	2.082
9	8.111	1.164	16.574	1.718	24.883	2.080
10	8.989	1.160	18.390	1.716	27.621	2.077
12	10.744	1.155	22.022	1.712	33.098	2.074
14	12.499	1.151	25.653	1.709	38.574	2.072
16	14.252	1.148	29.284	1.707	44.051	2.070
18	16.006	1.146	32.914	1.705	49.527	2.069
20	17.759	1.144	36.545	1.704	55.003	2.068
	n=5		n=6		n=7	
1	3.827	2.481	4.677	2.673	5.484	2.830
2	7.471	2.405	9.161	2.604	10.767	2.768
3	11.102	2.379	13.634	2.581	16.040	2.747
4	14.729	2.366	18.103	2.570	21.311	2.736
5	18.354	2.358	22.570	2.563	26.580	2.730
6	21.979	2.353	27.037	2.558	31.848	2.726
7	25.603	2.349	31.504	2.555	37.117	2.723
8	29.227	2.346	35.971	2.552	42.385	2.720
9	32.850	2.344	40.437	2.550	47.652	2.719
10	36.474	2.342	44.903	2.549	52.920	2.717
12	43.720	2.339	53.835	2.546	63.455	2.715
14	50.967	2.337	62.767	2.545	73.991	2.714
16	58.213	2.336	71.699	2.543	84.526	2.712
18	65.459	2.335	80.631	2.542	95.061	2.711
20	72.705	2.334	89.562	2.541	105.596	2.711
	n=8		n=9		n=10	
1	6.251	2.963	6.982	3.078	7.680	3.179
2	12.296	2.906	13.753	3.024	15.146	3.129
3	18.331	2.886	20.516	3.006	22.604	3.112
4	24.365	2.877	27.277	2.997	30.060	3.103
5	30.397	2.871	34.037	2.992	37.516	3.098
6	36.428	2.867	40.796	2.988	44.970	3.096
7	42.459	2.864	47.555	2.986	52.425	3.092
8	48.491	2.862	54.314	2.984	59.880	3.090
9	54.522	2.860	61.073	2.982	67.334	3.089
10	60.552	2.859	67.832	2.981	74.789	3.088
12	72.614	2.857	81.349	2.979	89.697	3.086
14	84.675	2.856	94.866	2.978	104.606	3.085
16	96.737	2.855	108.383	2.977	119.514	3.084
18	108.798	2.854	121.900	2.976	134.423	3.083
20	120.859	2.853	135.417	2.976	149.331	3.083

Table 17 FACTORS FOR CONTROL CHARTS (Table: p. 43)

This table gives factors for setting up 3σ control charts for the mean, range and standard deviation of a variable. The sampling is to be done with a sample of size n (the integer, given in column one of the table). From past experience let $\bar{\bar{X}}$, $\bar{R}$ and $\bar{\sigma}$ be "stable" estimates of the mean, range and standard deviation based upon samples of size n when the process is in control. The quantities $\bar{\bar{X}}$, $\bar{R}$ and $\bar{\sigma}$ are the averages of past sample means, ranges and standard deviations, $\hat{\sigma}$, defined as

$$\hat{\sigma} = \sqrt{\frac{\sum_{i=1}^{n}(X_i - \bar{X})^2}{n}}$$

Let s be the usual sample standard deviation; that is,

$$s = \sqrt{\frac{n}{(n-1)}}\,\hat{\sigma}$$

The control charts are constructed as:

Chart	Central Line	3 Control Limits
$\bar{X}$	$\bar{\bar{X}}$	$\bar{\bar{X}} \pm A_1\bar{\sigma}$
		or $\bar{\bar{X}} \pm \sqrt{\frac{(n-1)}{n}}\,A_1\bar{s}$
		or $\bar{\bar{X}} \pm A_2\bar{R}$
R	$\bar{R}$	$D_3\bar{R}$ and $D_4\bar{R}$
s	$\bar{s} = \sqrt{\frac{n}{(n-1)}}\,\bar{\sigma}$	$B_3\bar{s}$ and $B_4\bar{s}$
$\hat{\sigma}$	$\bar{\sigma}$	$B_3\bar{\sigma}$ and $B_4\bar{\sigma}$

where A_1, A_2, D_3, D_4, B_3 and B_4 are given in columns as labeled for given n.

Table 17 FACTORS FOR CONTROL CHARTS
(Description: p. 42)

n	A_1	A_2	D_3	D_4	B_3	B_4
2	3.760	1.880	0	3.267	0	3.267
3	2.394	1.023	0	2.575	0	2.568
4	1.880	0.729	0	2.282	0	2.266
5	1.596	0.577	0	2.114	0	2.089
6	1.410	0.483	0	2.004	0.030	1.970
7	1.277	0.419	0.076	1.924	0.118	1.882
8	1.175	0.373	0.136	1.864	0.185	1.815
9	1.094	0.337	0.184	1.816	0.239	1.761
10	1.028	0.308	0.223	1.777	0.284	1.716
11	0.973	0.285	0.256	1.744	0.321	1.679
12	0.925	0.266	0.283	1.717	0.354	1.646
13	0.884	0.249	0.307	1.693	0.382	1.618
14	0.848	0.235	0.328	1.672	0.406	1.594
15	0.816	0.223	0.347	1.653	0.428	1.572
16	0.788	0.212	0.363	1.637	0.448	1.552
17	0.762	0.203	0.378	1.622	0.466	1.534
18	0.738	0.194	0.391	1.609	0.482	1.518
19	0.717	0.187	0.404	1.596	0.497	1.503
20	0.697	0.180	0.415	1.585	0.510	1.490
21	0.679	0.173	0.425	1.575	0.523	1.477
22	0.662	0.167	0.435	1.565	0.534	1.466
23	0.647	0.162	0.443	1.557	0.545	1.455
24	0.632	0.157	0.452	1.548	0.555	1.445
25	0.619	0.153	0.459	1.541	0.565	1.435

Table 18 CHARACTERISTICS OF SEQUENTIAL TESTS OF A BINOMIAL PROPORTION (Table: p. 45)

Table 18 contains values for using Wald's sequential probability ratio test in a Bernoulli situation for testing p_1 versus p_2, $p_1 < p_2$, with $\alpha = 0.05$ and $\beta = 0.10$. The larger proportion p_2 is accepted and testing is stopped when the number of successes in n trials is greater than or equal to $h_2 + sn$. The smaller proportion p_1 is accepted and testing is stopped if the number of successes in n trials is less than or equal to $-h_1 + sn$. The last five columns give the expected number of observations $E(n|p)$ where p is the true proportion of successes, for $p = 0$, $p = 1$, $p = p_1$, $p = s$ and $p = p_2$.

Table 19 CRITICAL VALUES FOR TESTING AN EXTREME VALUE (Table: pp. 46-47)

Let $X_1, \ldots, X_n$ be a sample from a normal distribution with mean zero and variance σ^2, which are independent of $(\nu s^2/\sigma^2)$, which has a chi square distribution with ν degrees of freedom. Let

$$T = \max_{i=1,\ldots,n} \frac{|X_i|}{s}$$

For given ν, n and cumulative probability, γ, the critical values t are tabulated such that

$$P(T \leq t) = \gamma$$

The statistic is used in setting up simultaneous confidence intervals for orthogonal estimates in the analysis of variance as well as in testing extreme values.

Table 18 CHARACTERISTICS OF SEQUENTIAL TESTS OF A BINOMIAL PROPORTION (Description: p. 44)

p_2	h_2	h_1	s	$E(n\|0)$	$E(n\|1)$	$E(n\|p_1)$	$E(n\|s)$	$E(n\|p_2)$
(p_1 = 0.005)								
0.01	4.1398	3.2245	0.00722	447	5	1289	1863	1222
0.02	2.0624	1.6064	0.01084	149	3	244	309	185
0.03	1.5906	1.2369	0.01400	89	2	122	143	82
0.04	1.3664	1.0643	0.01693	63	2	79	87	49
0.05	1.2305	0.9585	0.01970	49	2	58	61	33
0.06	1.1371	0.8857	0.02237	40	2	45	46	25
0.07	1.0679	0.8318	0.02496	34	2	37	36	19
(p_1 = 0.010)								
0.03	2.5829	2.0118	0.01824	111	3	216	290	181
0.04	2.0397	1.5887	0.02172	74	3	120	153	92
0.05	1.7510	1.3639	0.02499	55	2	81	98	58
0.06	1.5678	1.2211	0.02811	44	2	60	70	40
0.07	1.4391	1.1209	0.03113	37	2	47	53	30
0.08	1.3426	1.0458	0.03406	31	2	38	43	24
(p_1 = 0.015)								
0.03	4.0796	3.1776	0.02166	147	5	423	612	402
0.04	2.8716	2.2367	0.02654	88	3	188	258	163
0.05	2.3307	1.8153	0.02917	63	3	113	149	92
0.06	2.0169	1.5710	0.03263	49	3	79	100	61
0.07	1.8089	1.4089	0.03596	40	2	60	74	44
(p_1 = 0.020)								
0.03	6.9527	5.4154	0.02467	220	8	1027	1565	1073
0.04	4.0495	3.1541	0.02889	110	5	314	455	300
0.05	3.0509	2.3763	0.03282	73	4	164	228	146
0.06	2.5348	1.9743	0.03655	55	3	106	142	89
0.07	2.2146	1.7250	0.04012	43	3	76	99	61
0.08	1.9941	1.5532	0.04359	36	3	58	74	45
0.09	1.8315	1.4265	0.04696	31	2	47	58	35
0.10	1.7056	1.3285	0.05025	27	2	39	47	28

Table 19 CRITICAL VALUES FOR TESTING AN EXTREME VALUE (Description: p. 44)

cumulative probability, γ = 0.90

ν	n=3	n=4	n=5	n=6	n=7	n=8	n=9	n=10
4	2.98	3.20	3.37	3.51	3.62	3.72	3.81	3.89
6	2.64	2.82	2.96	3.07	3.17	3.25	3.32	3.38
8	2.49	2.66	2.78	2.88	2.96	3.04	3.10	3.16
10	2.41	2.56	2.68	2.77	2.85	2.92	2.98	3.03
12	2.36	2.50	2.61	2.70	2.78	2.84	2.90	2.95
14	2.32	2.46	2.57	2.65	2.72	2.79	2.84	2.89
16	2.29	2.43	2.53	2.62	2.69	2.75	2.80	2.85
18	2.27	2.40	2.51	2.59	2.66	2.72	2.77	2.81
20	2.26	2.39	2.49	2.57	2.63	2.69	2.74	2.79
24	2.23	2.36	2.46	2.53	2.60	2.65	2.70	2.75
30	2.21	2.33	2.43	2.50	2.57	2.62	2.67	2.71
40	2.18	2.30	2.40	2.47	2.53	2.58	2.63	2.67
60	2.16	2.28	2.37	2.44	2.50	2.55	2.59	2.63
120	2.14	2.25	2.34	2.41	2.47	2.52	2.56	2.60
∞	2.11	2.23	2.31	2.38	2.43	2.48	2.52	2.56

cumulative probability, γ = 0.95

ν	n=3	n=4	n=5	n=6	n=7	n=8	n=9	n=10
4	3.74	4.00	4.20	4.37	4.50	4.62	4.72	4.82
6	3.19	3.39	3.54	3.66	3.77	3.86	3.94	4.01
8	2.96	3.13	3.26	3.36	3.45	3.53	3.60	3.66
10	2.83	2.98	3.10	3.20	3.28	3.35	3.41	3.47
12	2.75	2.89	3.00	3.09	3.17	3.24	3.29	3.34
14	2.69	2.83	2.94	3.02	3.09	3.16	3.21	3.26
16	2.65	2.78	2.89	2.97	3.04	3.10	3.15	3.20
18	2.62	2.75	2.85	2.93	3.00	3.05	3.11	3.15
20	2.59	2.72	2.82	2.90	2.96	3.02	3.07	3.11
24	2.56	2.68	2.77	2.85	2.91	2.97	3.02	3.06
30	2.52	2.64	2.73	2.80	2.87	2.92	2.96	3.00
60	2.45	2.56	2.65	2.72	2.77	2.82	2.86	2.90
120	2.42	2.53	2.61	2.67	2.73	2.77	2.81	2.85
∞	2.39	2.49	2.57	2.63	2.68	2.73	2.77	2.80

Table 19 CRITICAL VALUES FOR TESTING
VALUE (Description: p. 44

cumulative probability, γ =

ν	n=3	n=4	n=5	n=6	n=7	
4	4.62	4.92	5.16	5.35	5.51	[illegible]
6	3.77	3.99	4.15	4.29	4.41	[illegible]
8	3.43	3.61	3.75	3.86	3.95	4
10	3.24	3.40	3.53	3.63	3.71	3
12	3.13	3.27	3.39	3.48	3.56	3.
14	3.05	3.19	3.29	3.38	3.45	3.
16	2.99	3.12	3.23	3.31	3.38	3.4
18	2.95	3.08	3.18	3.25	3.32	3.3
20	2.91	3.04	3.13	3.21	3.28	3.3
24	2.86	2.98	3.08	3.15	3.21	3.27
30	2.82	2.93	3.02	3.09	3.15	3.20
40	2.77	2.88	2.96	3.03	3.09	3.13
60	2.72	2.83	2.91	2.97	3.03	3.07
120	2.68	2.78	2.85	2.92	2.97	3.01
∞	2.64	2.73	2.80	2.86	2.91	2.95

cumulative probability, γ = 0.99

ν	n=3	n=4	n=5	n=6	n=7	n=8	n=	
4	5.98	6.36	6.65	6.89	7.10	7.27	7.4	
6	4.61	4.85	5.04	5.20	5.33	5.45	5.5	
8	4.08	4.27	4.42	4.55	4.65	4.74	4.8	
10	3.80	3.97	4.10	4.20	4.29	4.37	4.44	
12	3.63	3.78	3.90	3.99	4.07	4.14	4.20	
14	3.52	3.66	3.76	3.85	3.93	3.99	4.05	
16	3.43	3.57	3.67	3.75	3.82	3.88	3.94	
18	3.37	3.50	3.60	3.67	3.74	3.80	3.85	
20	3.32	3.44	3.54	3.62	3.68	3.74	3.78	
24	3.25	3.37	3.46	3.53	3.59	3.64	3.69	
30	3.18	3.29	3.38	3.45	3.50	3.55	3.60	[illegible]
40	3.12	3.22	3.30	3.37	3.42	3.47	3.51	3
60	3.05	3.15	3.23	3.29	3.34	3.38	3.42	3
120	2.99	3.09	3.16	3.21	3.26	3.30	3.34	3.
∞	2.93	3.02	3.09	3.14	3.19	3.23	3.26	3.

Table 2

Milita
Tables
plans
Unite
are t
batch
spec
acce
leve
cen
hun
nor
the
fo
fo
sa
o

0 MIL-STD 105D OR ABC PLAN: SINGLE SAMPLING (Tables: pp. 50-53)

ry Standard 105D, Sampling Procedures and for Inspection by Attributes, gives sampling developed by an international committee from the d States, Great Britain, and Canada. The tables o be used for a continuing series of lots or es. To use the tables the supplier and consumer ify the lot or batch size to be inspected and an ptable quality level, AQL, as well as an inspection l. The AQL is defined either in terms of the per- t (not fraction) of items defective or defects per dred units. Inspection is at one of three levels: rmal, tightened, or reduced, usually beginning with e normal level and switching between plans as in the llowing description. This pocketbook contains tables r the single sampling plan. Double and multiple mpling plans are also available in the Department f Defense document.

To find the sampling scheme use Table 20A with the appropriate batch size to find a sample size code letter (column I refers to a loose inspection level, column II is normally used, and column III gives greater discrimination). Find the row corresponding to the sample size code letter and the column corresponding to the AQL in Table 20A, 20B, or 20C, depending upon whether normal, tightened, or reduced sampling is being used. Sampling is usually begun on the normal sampling table. If the intersection of the row and column contains two numbers, the first number is the acceptance number and the second the rejection number. The sample is accepted if the percent defective or number of defects (depending upon the definition of the AQL) is

less than or equal to this number. If the row and column intersect on an arrow, the arrow should be followed in the given column to the acceptance and rejection numbers to which it points. The sample size of the new row is to be used. If the sample size exceeds the lot size, 100% inspection is to be done.

If two out of five consecutive lots fail at a normal sampling level the tightened procedure is instituted. If five consecutive tightened procedures are acceptable the normal procedure is reinstituted. If ten consecutive lots remain on tightened inspection the plan is terminated pending appropriate action. Reduced inspection is instituted if four things occur: 1) the preceeding ten batches have been accepted, 2) the total number of defectives (or defects) in the preceeding ten samples is less than or equal to the number in Table VIII (not included here), 3) the production is at a steady rate, and 4) reduced inspection is considered desireable by the responsible authority. Inspection is returned to normal from reduced if 1) a lot is rejected, 2) production becomes irregular or delayed, or 3) other conditions warrant that normal inspection be instituted.

This brief introduction does not include such topics as 1) critical, major, and minor defects, 2) double and multiple sampling, 3) average sample size curves, and 4) operating characteristic curves, all of which are considered in Military Standard 105D.

Table 20A MIL-STD 105D OR ABC PLAN: SAMPLE SIZE CODE LETTERS (Description: pp. 48-49)

Lot or batch size	Special inspection levels				General inspection levels		
	S-1	S-2	S-3	S-4	I	II	III
2 to 8	A	A	A	A	A	A	B
9 to 15	A	A	A	A	A	B	C
16 to 25	A	A	B	B	B	C	D
26 to 50	A	B	B	C	C	D	E
51 to 90	B	B	C	C	C	E	F
91 to 150	B	B	C	D	D	F	G
151 to 280	B	C	D	E	E	G	H
281 to 500	B	C	D	E	F	H	J
501 to 1200	C	C	E	F	G	J	K
1201 to 3200	C	D	E	G	H	K	L
3201 to 10000	C	D	F	G	J	L	M
10001 to 35000	C	D	F	H	K	M	N
35001 to 150000	D	E	G	J	L	N	P
150001 to 500000	D	E	G	J	M	P	Q
500001 and over	D	E	H	K	N	Q	R

Table 20B MIL-STD 105D OR ABC PLAN: SINGLE SAMPLING (Description: pp. 48-49)

Sample size code letter	Sample size	Acceptable Quality Levels (normal inspection)																									
		0.010	0.015	0.025	0.040	0.065	0.10	0.15	0.25	0.40	0.65	1.0	1.5	2.5	4.0	6.5	10	15	25	40	65	100	150	250	400	650	1000
		Ac Re	Ac Re	Ac Re	Ac Re	Ac Re	Ac Re	Ac Re	Ac Re	Ac Re	Ac Re	Ac Re	Ac Re	Ac Re	Ac Re	Ac Re	Ac Re	Ac Re	Ac Re	Ac Re	Ac Re	Ac Re	Ac Re	Ac Re	Ac Re	Ac Re	Ac Re
A	2	↓	↓	↓	↓	↓	↓	↓	↓	↓	↓	↓	↓	↓	↓	0 1	↓	↓	1 2	2 3	3 4	5 6	7 8	10 11	14 15	21 22	30 31
B	3	↓	↓	↓	↓	↓	↓	↓	↓	↓	↓	↓	↓	↓	0 1	↑	↓	1 2	2 3	3 4	5 6	7 8	10 11	14 15	21 22	30 31	44 45
C	5	↓	↓	↓	↓	↓	↓	↓	↓	↓	↓	↓	↓	0 1	↑	↓	1 2	2 3	3 4	5 6	7 8	10 11	14 15	21 22	30 31	44 45	↑
D	8	↓	↓	↓	↓	↓	↓	↓	↓	↓	↓	↓	0 1	↑	↓	1 2	2 3	3 4	5 6	7 8	10 11	14 15	21 22	30 31	44 45	↑	↑
E	13	↓	↓	↓	↓	↓	↓	↓	↓	↓	↓	0 1	↑	↓	1 2	2 3	3 4	5 6	7 8	10 11	14 15	21 22	30 31	44 45	↑	↑	↑
F	20	↓	↓	↓	↓	↓	↓	↓	↓	↓	0 1	↑	↓	1 2	2 3	3 4	5 6	7 8	10 11	14 15	21 22	↑	↑	↑	↑	↑	↑
G	32	↓	↓	↓	↓	↓	↓	↓	↓	0 1	↑	↓	1 2	2 3	3 4	5 6	7 8	10 11	14 15	21 22	↑	↑	↑	↑	↑	↑	↑
H	50	↓	↓	↓	↓	↓	↓	↓	0 1	↑	↓	1 2	2 3	3 4	5 6	7 8	10 11	14 15	21 22	↑	↑	↑	↑	↑	↑	↑	↑
J	80	↓	↓	↓	↓	↓	↓	0 1	↑	↓	1 2	2 3	3 4	5 6	7 8	10 11	14 15	21 22	↑	↑	↑	↑	↑	↑	↑	↑	↑
K	125	↓	↓	↓	↓	↓	0 1	↑	↓	1 2	2 3	3 4	5 6	7 8	10 11	14 15	21 22	↑	↑	↑	↑	↑	↑	↑	↑	↑	↑
L	200	↓	↓	↓	↓	0 1	↑	↓	1 2	2 3	3 4	5 6	7 8	10 11	14 15	21 22	↑	↑	↑	↑	↑	↑	↑	↑	↑	↑	↑
M	315	↓	↓	↓	0 1	↑	↓	1 2	2 3	3 4	5 6	7 8	10 11	14 15	21 22	↑	↑	↑	↑	↑	↑	↑	↑	↑	↑	↑	↑
N	500	↓	↓	0 1	↑	↓	1 2	2 3	3 4	5 6	7 8	10 11	14 15	21 22	↑	↑	↑	↑	↑	↑	↑	↑	↑	↑	↑	↑	↑
P	800	↓	0 1	↑	↓	1 2	2 3	3 4	5 6	7 8	10 11	14 15	21 22	↑	↑	↑	↑	↑	↑	↑	↑	↑	↑	↑	↑	↑	↑
Q	1250	0 1	↑	↓	1 2	2 3	3 4	5 6	7 8	10 11	14 15	21 22	↑	↑	↑	↑	↑	↑	↑	↑	↑	↑	↑	↑	↑	↑	↑
R	2000	↑	↑	1 2	2 3	3 4	5 6	7 8	10 11	14 15	21 22	↑	↑	↑	↑	↑	↑	↑	↑	↑	↑	↑	↑	↑	↑	↑	↑

↓ = Use first sampling plan below arrow. If sample size equals, or exceeds, lot or batch size, do 100 percent inspection.

↑ = Use first sampling plan above arrow.

Ac = Acceptance number.

Re = Rejection number.

Table 20C MIL-STD 105D OR ABC PLAN: SINGLE SAMPLING (Description: pp. 48-49)

Sample size code letter	Sample size	Acceptable Quality Levels (tightened inspection)																									
		0.010	0.015	0.025	0.040	0.065	0.10	0.15	0.25	0.40	0.65	1.0	1.5	2.5	4.0	6.5	10	15	25	40	65	100	150	250	400	650	1000
		Ac Re	Ac Re	Ac Re	Ac Re	Ac Re	Ac Re	Ac Re	Ac Re	Ac Re	Ac Re	Ac Re	Ac Re	Ac Re	Ac Re	Ac Re	Ac Re	Ac Re	Ac Re	Ac Re	Ac Re	Ac Re	Ac Re	Ac Re	Ac Re	Ac Re	Ac Re
A	2	⇩	⇩	⇩	⇩	⇩	⇩	⇩	⇩	⇩	⇩	⇩	⇩	⇩	⇩	⇩	⇩	⇩	⇩	1 2	2 3	3 4	5 6	8 9	12 13	18 19	27 28
B	3	⇩	⇩	⇩	⇩	⇩	⇩	⇩	⇩	⇩	⇩	⇩	⇩	⇩	⇩	0 1	⇩	⇩	1 2	2 3	3 4	5 6	8 9	12 13	18 19	27 28	41 42
C	5	⇩	⇩	⇩	⇩	⇩	⇩	⇩	⇩	⇩	⇩	⇩	⇩	⇩	0 1	⇩	⇩	1 2	2 3	3 4	5 6	8 9	12 13	18 19	27 28	41 42	⇧
D	8	⇩	⇩	⇩	⇩	⇩	⇩	⇩	⇩	⇩	⇩	⇩	⇩	0 1	⇩	⇩	1 2	2 3	3 4	5 6	8 9	12 13	18 19	27 28	41 42	⇧	⇧
E	13	⇩	⇩	⇩	⇩	⇩	⇩	⇩	⇩	⇩	⇩	⇩	0 1	⇩	⇩	1 2	2 3	3 4	5 6	8 9	12 13	18 19	27 28	41 42	⇧	⇧	⇧
F	20	⇩	⇩	⇩	⇩	⇩	⇩	⇩	⇩	⇩	⇩	0 1	⇩	⇩	1 2	2 3	3 4	5 6	8 9	12 13	18 19	⇧	⇧	⇧	⇧	⇧	⇧
G	32	⇩	⇩	⇩	⇩	⇩	⇩	⇩	⇩	⇩	0 1	⇩	⇩	1 2	2 3	3 4	5 6	8 9	12 13	18 19	⇧	⇧	⇧	⇧	⇧	⇧	⇧
H	50	⇩	⇩	⇩	⇩	⇩	⇩	⇩	⇩	0 1	⇩	⇩	1 2	2 3	3 4	5 6	8 9	12 13	18 19	⇧	⇧	⇧	⇧	⇧	⇧	⇧	⇧
J	80	⇩	⇩	⇩	⇩	⇩	⇩	⇩	0 1	⇩	⇩	1 2	2 3	3 4	5 6	8 9	12 13	18 19	⇧	⇧	⇧	⇧	⇧	⇧	⇧	⇧	⇧
K	125	⇩	⇩	⇩	⇩	⇩	⇩	0 1	⇩	⇩	1 2	2 3	3 4	5 6	8 9	12 13	18 19	⇧	⇧	⇧	⇧	⇧	⇧	⇧	⇧	⇧	⇧
L	200	⇩	⇩	⇩	⇩	⇩	0 1	⇩	⇩	1 2	2 3	3 4	5 6	8 9	12 13	18 19	⇧	⇧	⇧	⇧	⇧	⇧	⇧	⇧	⇧	⇧	⇧
M	315	⇩	⇩	⇩	⇩	0 1	⇩	⇩	1 2	2 3	3 4	5 6	8 9	12 13	18 19	⇧	⇧	⇧	⇧	⇧	⇧	⇧	⇧	⇧	⇧	⇧	⇧
N	500	⇩	⇩	⇩	0 1	⇩	⇩	1 2	2 3	3 4	5 6	8 9	12 13	18 19	⇧	⇧	⇧	⇧	⇧	⇧	⇧	⇧	⇧	⇧	⇧	⇧	⇧
P	800	⇩	⇩	0 1	⇩	⇩	1 2	2 3	3 4	5 6	8 9	12 13	18 19	⇧	⇧	⇧	⇧	⇧	⇧	⇧	⇧	⇧	⇧	⇧	⇧	⇧	⇧
Q	1250	⇩	0 1	⇩	⇩	1 2	2 3	3 4	5 6	8 9	12 13	18 19	⇧	⇧	⇧	⇧	⇧	⇧	⇧	⇧	⇧	⇧	⇧	⇧	⇧	⇧	⇧
R	2000	0 1	⇧	⇩	1 2	2 3	3 4	5 6	8 9	12 13	18 19	⇧	⇧	⇧	⇧	⇧	⇧	⇧	⇧	⇧	⇧	⇧	⇧	⇧	⇧	⇧	⇧
S	3150			1 2																							

⇩ = Use first sampling plan below arrow. If sample size equals or exceeds lot or batch size, do 100 percent inspection.
⇧ = Use first sampling plan above arrow.

Ac = Acceptance number.
Re = Rejection number.

Table 20D MIL-STD 105D OR ABC PLAN: SINGLE SAMPLING (Description: pp. 48-49)

Sample size code letter	Sample size	Acceptable Quality Levels (reduced inspection)†																									
		0.010	0.015	0.025	0.040	0.065	0.10	0.15	0.25	0.40	0.65	1.0	1.5	2.5	4.0	6.5	10	15	25	40	65	100	150	250	400	650	1000
		Ac Re	Ac Re	Ac Re	Ac Re	Ac Re	Ac Re	Ac Re	Ac Re	Ac Re	Ac Re	Ac Re	Ac Re	Ac Re	Ac Re	Ac Re	Ac Re	Ac Re	Ac Re	Ac Re	Ac Re	Ac Re	Ac Re	Ac Re	Ac Re	Ac Re	Ac Re
A	2	↓	↓	↓	↓	↓	↓	↓	↓	↓	↓	↓	↓	↓	↓	0 1	↓	↓	1 2	2 3	3 4	5 6	7 8	10 11	14 15	21 22	30 31
B	2	↓	↓	↓	↓	↓	↓	↓	↓	↓	↓	↓	↓	↓	0 1	↑	↓	0 2	1 3	2 4	3 5	5 6	7 8	10 11	14 15	21 22	30 31
C	2	↓	↓	↓	↓	↓	↓	↓	↓	↓	↓	↓	↓	0 1	↑	↓	0 2	1 3	1 4	2 5	3 6	5 8	7 10	10 13	14 17	21 24	↑
D	3	↓	↓	↓	↓	↓	↓	↓	↓	↓	↓	↓	0 1	↑	↓	0 2	1 3	1 4	2 5	3 6	5 8	7 10	10 13	14 17	21 24	↑	↑
E	5	↓	↓	↓	↓	↓	↓	↓	↓	↓	↓	0 1	↑	↓	0 2	1 3	1 4	2 5	3 6	5 8	7 10	10 13	14 17	21 24	↑	↑	↑
F	8	↓	↓	↓	↓	↓	↓	↓	↓	↓	0 1	↑	↓	0 2	1 3	1 4	2 5	3 6	5 8	7 10	10 13	↑	↑	↑	↑	↑	↑
G	13	↓	↓	↓	↓	↓	↓	↓	↓	0 1	↑	↓	0 2	1 3	1 4	2 5	3 6	5 8	7 10	10 13	↑	↑	↑	↑	↑	↑	↑
H	20	↓	↓	↓	↓	↓	↓	↓	0 1	↑	↓	0 2	1 3	1 4	2 5	3 6	5 8	7 10	10 13	↑	↑	↑	↑	↑	↑	↑	↑
J	32	↓	↓	↓	↓	↓	↓	0 1	↑	↓	0 2	1 3	1 4	2 5	3 6	5 8	7 10	10 13	↑	↑	↑	↑	↑	↑	↑	↑	↑
K	50	↓	↓	↓	↓	↓	0 1	↑	↓	0 2	1 3	1 4	2 5	3 6	5 8	7 10	10 13	↑	↑	↑	↑	↑	↑	↑	↑	↑	↑
L	80	↓	↓	↓	↓	0 1	↑	↓	0 2	1 3	1 4	2 5	3 6	5 8	7 10	10 13	↑	↑	↑	↑	↑	↑	↑	↑	↑	↑	↑
M	125	↓	↓	↓	0 1	↑	↓	0 2	1 3	1 4	2 5	3 6	5 8	7 10	10 13	↑	↑	↑	↑	↑	↑	↑	↑	↑	↑	↑	↑
N	200	↓	↓	0 1	↑	↓	0 2	1 3	1 4	2 5	3 6	5 8	7 10	10 13	↑	↑	↑	↑	↑	↑	↑	↑	↑	↑	↑	↑	↑
P	315	↓	0 1	↑	↓	0 2	1 3	1 4	2 5	3 6	5 8	7 10	10 13	↑	↑	↑	↑	↑	↑	↑	↑	↑	↑	↑	↑	↑	↑
Q	500	0 1	↑	↓	0 2	1 3	1 4	2 5	3 6	5 8	7 10	10 13	↑	↑	↑	↑	↑	↑	↑	↑	↑	↑	↑	↑	↑	↑	↑
R	800	↑	↑	0 2	1 3	1 4	2 5	3 6	5 8	7 10	10 13	↑	↑	↑	↑	↑	↑	↑	↑	↑	↑	↑	↑	↑	↑	↑	↑

↓ = Use first sampling plan below arrow. If sample size equals or exceeds lot or batch size, do 100 percent inspection.
↑ = Use first sampling plan above arrow.
Ac = Acceptance number.
Re = Rejection number.
† = If the acceptance number has been exceeded, but the rejection number has not been reached, accept the lot, but reinstate normal inspection (see 10.1.4).

Table 21 BINOMIAL ACCEPTANCE PLANS WITH FIXED α AND β
(Table: p. 55)

A binomial acceptance procedure is to be designed that accepts a lot or batch if c or fewer defectives are observed in a sample of size n. The probability of rejection must be less than or equal to α and $1 - \beta$ when the fractions defective are $p = p_1$ and p_2, respectively. To use the table, find the correct α, β column and locate the smallest entry in that column greater than or equal to p_2/p_1. The row of that entry gives c and np_1. The sample size n is found by dividing np_1 by p_1. Table 21 uses the Poisson approximation to the binomial.

Table 22 THE CUMULATIVE POISSON DISTRIBUTION
(Table: pp. 56-57)

Tabulated are values of $\sum_{j=0}^{k} \lambda^j e^{-\lambda}/j! = P$ for given k and λ.

Table 23 CONFIDENCE LIMITS ON A POISSON PARAMETER
(Table: pp. 58-59)

For given integer $c \geq 0$ and probability P, Table 23 contains the value λ such that $\sum_{j=0}^{c} \lambda^j e^{-\lambda}/j! = P$. The table allows calculation of one- and two-sided confidence limits for λ. Example: Observe c = 4, find the 95% confidence interval of the form a) $(0, \lambda)$; b) (λ, ∞); c) (λ_L, λ_u). Answer: a) enter under the 0.95 column (0, 9.154); b) enter under the 0.05 column (1.970, ∞); c) enter under the 0.025 and 0.0975 columns (1.623, 10.242).

Table 21 BINOMIAL ACCEPTANCE PLANS WITH FIXED α AND β
(Description: p. 54)

c	p_2/p_1 α=.05 β=.10	p_2/p_1 α=.05 β=.05	np_1	c	p_2/p_1 α=.01 β=.10	p_2/p_1 α=.01 β=.05	np_1
0	44.891	58.404	0.051	0	229.105	298.073	0.010
1	10.946	13.349	0.355	1	26.184	31.933	0.149
2	6.509	7.699	0.818	2	12.206	14.438	0.436
3	4.890	5.675	1.366	3	8.115	9.418	0.823
4	4.057	4.646	1.970	4	6.249	7.156	1.279
5	3.549	4.023	2.613	5	5.195	5.889	1.785
6	3.206	3.605	3.285	6	4.520	5.082	2.330
7	2.957	3.303	3.981	7	4.050	4.524	2.906
8	2.768	3.074	4.695	8	3.705	4.115	3.507
9	2.618	2.895	5.425	9	3.440	3.803	4.130
10	2.497	2.750	6.169	10	3.229	3.555	4.771
11	2.397	2.630	6.924	11	3.058	3.354	5.428
12	2.312	2.528	7.690	12	2.915	3.188	6.099
13	2.240	2.442	8.464	13	2.795	3.047	6.782
14	2.177	2.367	9.246	14	2.692	2.927	7.477
15	2.122	2.301	10.036	15	2.603	2.823	8.181
16	2.073	2.243	10.832	16	2.524	2.732	8.895
17	2.029	2.192	11.634	17	2.455	2.652	9.616
18	1.990	2.145	12.442	18	2.393	2.580	10.346
19	1.954	2.103	13.255	19	2.337	2.516	11.082
20	1.922	2.065	14.072	20	2.287	2.458	11.825
21	1.892	2.030	14.894	21	2.241	2.405	12.574
22	1.865	1.998	15.719	22	2.200	2.357	13.329
23	1.840	1.969	16.549	23	2.162	2.313	14.089
24	1.817	1.942	17.382	24	2.126	2.272	14.853
25	1.795	1.917	18.219	25	2.094	2.235	15.623
26	1.775	1.893	19.058	26	2.064	2.200	16.397
27	1.757	1.871	19.901	27	2.036	2.168	17.175
28	1.739	1.850	20.746	28	2.009	2.138	17.957
29	1.723	1.831	21.594	29	1.985	2.110	18.742
30	1.707	1.813	22.445	30	1.962	2.083	19.532
31	1.692	1.796	23.297	31	1.940	2.059	20.324
32	1.679	1.780	24.153	32	1.920	2.035	21.120
33	1.665	1.764	25.010	33	1.900	2.013	21.919
34	1.653	1.750	25.870	34	1.882	1.992	22.721
35	1.641	1.736	26.731	35	1.865	1.973	23.526
36	1.630	1.723	27.595	36	1.848	1.954	24.333
37	1.619	1.710	28.460	37	1.833	1.936	25.143
38	1.609	1.698	29.327	38	1.818	1.919	25.955
39	1.599	1.687	30.196	39	1.804	1.903	26.770

Table 22 THE CUMULATIVE POISSON DISTRIBUTION
(Description: p. 54)

k\λ	0.001	0.005	0.010	0.015	0.020
0	0.9990 0050	0.9950 1248	0.9900 4983	0.9851 1194	0.980 199
1	0.9999 9950	0.9999 8754	0.9999 5033	0.9998 8862	0.999 803
2	1.0000 0000	0.9999 9998	0.9999 9983	0.9999 9944	0.999 999
3		1.0000 0000	1.0000 0000	1.0000 0000	1.000 000

k\λ	0.025	0.030	0.035	0.040	0.050
0	0.975 310	0.970 446	0.965 605	0.960 789	0.951 229
1	0.999 693	0.999 559	0.999 402	0.999 221	0.998 791
2	0.999 997	0.999 996	0.999 993	0.999 990	0.999 980
3	1.000 000	1.000 000	1.000 000	1.000 000	1.000 000

k\λ	0.060	0.070	0.075	0.080	0.090
0	0.941 765	0.932 394	0.927 743	0.923 116	0.913 931
1	0.998 270	0.997 661	0.997 324	0.996 966	0.996 185
2	0.999 966	0.999 946	0.999 934	0.999 920	0.999 886
3	0.999 999	0.999 999	0.999 999	0.999 998	0.999 997
4	1.000 000	1.000 000	1.000 000	1.000 000	1.000 000

k\λ	0.100	0.200	0.300	0.400	0.500
0	0.904 837	0.818 731	0.740 818	0.670 320	0.606 531
1	0.995 321	0.982 477	0.963 064	0.938 448	0.909 796
2	0.999 845	0.998 852	0.996 401	0.992 074	0.985 612
3	0.999 996	0.999 943	0.999 734	0.999 224	0.998 248
4	1.000 000	0.999 998	0.999 984	0.999 939	0.999 828
5		1.000 000	0.999 999	0.999 996	0.999 986
6			1.000 000	1.000 000	0.999 999
7					1.000 000

k\λ	0.600	0.700	0.800	0.900	1.000
0	0.548 812	0.496 585	0.449 329	0.406 570	0.367 879
1	0.878 099	0.844 195	0.808 792	0.772 482	0.735 759
2	0.976 885	0.965 858	0.952 577	0.937 143	0.919 699
3	0.996 642	0.994 247	0.990 920	0.986 541	0.981 012
4	0.999 606	0.999 214	0.998 589	0.997 656	0.996 340
5	0.999 961	0.999 910	0.999 816	0.999 657	0.999 406
6	0.999 997	0.999 991	0.999 979	0.999 957	0.999 917
7	1.000 000	0.999 999	0.999 998	0.999 995	0.999 990
8		1.000 000	1.000 000	1.000 000	0.999 999
9					1.000 000

Table 22 THE CUMULATIVE POISSON DISTRIBUTION
(Description: p. 54)

k\λ	1.20	1.40	1.60	1.80	2.00	2.50	3.00
0	0.3012	0.2466	0.2019	0.1653	0.1353	0.0821	0.0498
1	0.6626	0.5918	0.5249	0.4628	0.4060	0.2873	0.1991
2	0.8795	0.8335	0.7834	0.7306	0.6767	0.5438	0.4232
3	0.9662	0.9463	0.9212	0.8913	0.8571	0.7576	0.6472
4	0.9923	0.9857	0.9763	0.9636	0.9473	0.8912	0.8153
5	0.9985	0.9968	0.9940	0.9896	0.9834	0.9580	0.9161
6	0.9997	0.9994	0.9987	0.9974	0.9955	0.9858	0.9665
7	1.0000	0.9999	0.9997	0.9994	0.9989	0.9958	0.9881
8		1.0000	1.0000	0.9999	0.9998	0.9989	0.9962
9				1.0000	1.0000	0.9997	0.9989
10						0.9999	0.9997
11						1.0000	0.9999
12							1.0000

k\λ	4.00	5.00	6.00	7.00	8.00	9.00	10.00
0	0.0183	0.0067	0.0025	0.0009	0.0003	0.0001	0.0000
1	0.0916	0.0404	0.0174	0.0073	0.0030	0.0012	0.0005
2	0.2381	0.1247	0.0620	0.0296	0.0138	0.0062	0.0028
3	0.4335	0.2650	0.1512	0.0818	0.0424	0.0212	0.0103
4	0.6288	0.4405	0.2851	0.1730	0.0996	0.0550	0.0293
5	0.7851	0.6160	0.4457	0.3007	0.1912	0.1157	0.0671
6	0.8893	0.7622	0.6063	0.4497	0.3134	0.2068	0.1301
7	0.9489	0.8666	0.7440	0.5987	0.4530	0.3239	0.2202
8	0.9786	0.9319	0.8472	0.7291	0.5925	0.4557	0.3328
9	0.9919	0.9682	0.9161	0.8305	0.7166	0.5874	0.4579
10	0.9972	0.9863	0.9574	0.9015	0.8159	0.7060	0.5830
11	0.9991	0.9945	0.9799	0.9467	0.8881	0.8030	0.6968
12	0.9997	0.9980	0.9912	0.9730	0.9362	0.8758	0.7916
13	0.9999	0.9993	0.9964	0.9872	0.9658	0.9261	0.8645
14	1.0000	0.9998	0.9986	0.9943	0.9827	0.9585	0.9165
15		0.9999	0.9995	0.9976	0.9918	0.9780	0.9513
16		1.0000	0.9998	0.9990	0.9963	0.9889	0.9730
17			0.9999	0.9996	0.9984	0.9947	0.9857
18			1.0000	0.9999	0.9993	0.9976	0.9928
19				1.0000	0.9997	0.9989	0.9965
20					0.9999	0.9996	0.9984
21					1.0000	0.9998	0.9993
22						0.9999	0.9997
23						1.0000	0.9999
24							1.0000

Table 23 CONFIDENCE LIMITS ON A POISSON PARAMETER
(Description: p. 64)

	P					
c	0.005	0.010	0.025	0.050	0.100	0.250
0	0.005	0.010	0.025	0.051	0.105	0.288
1	0.103	0.149	0.242	0.355	0.532	0.961
2	0.338	0.436	0.619	0.818	1.102	1.727
3	0.672	0.823	1.090	1.366	1.745	2.535
4	1.078	1.279	1.623	1.970	2.433	3.369
5	1.537	1.785	2.202	2.613	3.152	4.219
6	2.037	2.330	2.814	3.285	3.895	5.083
7	2.571	2.906	3.454	3.981	4.656	5.956
8	3.132	3.507	4.115	4.695	5.432	6.838
9	3.717	4.130	4.795	5.425	6.221	7.726
10	4.321	4.771	5.491	6.169	7.021	8.620
11	4.943	5.428	6.201	6.924	7.829	9.519
12	5.580	6.099	6.922	7.690	8.646	10.422
13	6.231	6.782	7.654	8.464	9.470	11.329
14	6.893	7.477	8.395	9.246	10.300	12.239
15	7.567	8.181	9.145	10.036	11.135	13.152
16	8.251	8.895	9.903	10.832	11.976	14.068
17	8.943	9.616	10.668	11.634	12.822	14.987
18	9.644	10.346	11.439	12.442	13.671	15.907
19	10.353	11.082	12.217	13.255	14.525	16.830
20	11.069	11.825	12.999	14.072	15.383	17.755
21	11.792	12.574	13.787	14.894	16.244	18.682
22	12.521	13.329	14.580	15.719	17.108	19.610
23	13.255	14.089	15.377	16.549	17.975	20.540
24	13.995	14.853	16.179	17.382	18.844	21.471
25	14.741	15.623	16.984	18.219	19.717	22.404
26	15.491	16.397	17.793	19.058	20.592	23.338
27	16.245	17.175	18.606	19.901	21.469	24.273
28	17.004	17.957	19.422	20.746	22.348	25.209
29	17.767	18.742	20.241	21.594	23.229	26.147
30	18.534	19.532	21.063	22.445	24.113	27.085
31	19.305	20.324	21.888	23.297	24.998	28.025
32	20.079	21.120	22.716	24.153	25.885	28.965
33	20.857	21.919	23.546	25.010	26.774	29.907
34	21.638	22.721	24.379	25.870	27.664	30.849
35	22.422	23.526	25.214	26.731	28.556	31.792
36	23.208	24.333	26.051	27.595	29.450	32.736
37	23.998	25.143	26.891	28.460	30.345	33.681
38	24.791	25.955	27.733	29.327	31.241	34.626
39	25.586	26.770	28.577	30.196	32.139	35.572

Table 23 CONFIDENCE LIMITS ON A POISSON PARAMETER
(Description: p. 54)

c	P 0.750	0.900	0.950	0.975	0.990	0.995
0	1.386	2.303	2.996	3.689	4.605	5.298
1	2.693	3.890	4.744	5.572	6.638	7.430
2	3.920	5.322	6.296	7.225	8.406	9.274
3	5.109	6.681	7.754	8.767	10.045	10.977
4	6.274	7.994	9.154	10.242	11.605	12.594
5	7.423	9.275	10.513	11.668	13.108	14.150
6	8.558	10.532	11.842	13.059	14.571	15.660
7	9.684	11.771	13.148	14.423	16.000	17.134
8	10.802	12.995	14.435	15.763	17.403	18.578
9	11.914	14.206	15.705	17.085	18.783	19.998
10	13.020	15.407	16.962	18.390	20.145	21.398
11	14.121	16.598	18.208	19.682	21.490	22.779
12	15.217	17.782	19.443	20.962	22.821	24.145
13	16.310	18.958	20.669	22.230	24.139	25.497
14	17.400	20.128	21.886	23.490	25.446	26.836
15	18.486	21.292	23.097	24.740	26.743	28.164
16	19.570	22.452	24.301	25.983	28.030	29.482
17	20.652	23.606	25.499	27.219	29.310	30.791
18	21.731	24.756	26.692	28.448	30.581	32.091
19	22.808	25.903	27.879	29.671	31.845	33.383
20	23.883	27.045	29.062	30.888	33.103	34.668
21	24.956	28.184	30.240	32.101	34.355	35.946
22	26.028	29.320	31.415	33.308	35.601	37.218
23	27.098	30.453	32.585	34.511	36.841	38.484
24	28.167	31.584	33.752	35.710	38.077	39.745
25	29.234	32.711	34.916	36.905	39.308	41.000
26	30.300	33.836	36.077	38.096	40.534	42.251
27	31.365	34.959	37.234	39.284	41.757	43.497
28	32.428	36.080	38.389	40.468	42.975	44.738
29	33.491	37.199	39.541	41.649	44.190	45.976
30	34.552	38.315	40.691	42.827	45.401	47.209
31	35.613	39.430	41.838	44.002	46.608	48.439
32	36.672	40.543	42.982	45.174	47.813	49.665
33	37.731	41.654	44.125	46.344	49.014	50.888
34	38.788	42.764	45.266	47.512	50.213	52.107
35	39.845	43.872	46.404	48.677	51.408	53.324
36	40.901	44.978	47.541	49.839	52.601	54.537
37	41.957	46.083	48.675	51.000	53.791	55.748
38	43.011	47.187	49.808	52.158	54.979	56.955
39	44.065	48.289	50.940	53.314	56.164	58.161

Table 24 THE CUMULATIVE BINOMIAL DISTRIBUTION

(Table: pp. 61-69)

The quantity tabulated is

$$B(n,\ k,\ p) = \sum_{j=0}^{k} \frac{n!}{j!(n - j)!} p^{j}(1 - p)^{n-j}$$

For $p > 0.5$, the values of $B(n, k, p)$ can be obtained by using the relationship $B(n, k, p) = 1 - B(n, n + 1 - k, 1 - p)$. Values of the cumulative negative binomial distribution can be computed as follows:

$$p^{n} \sum_{j=0}^{k} \binom{n + j - 1}{n - 1} (1 - p)^{j} = 1 - B(k + n,\ n - 1,\ p)$$

Table 24 THE CUMULATIVE BINOMIAL DISTRIBUTION
(Description: p. 60)

		binomial probability, p					
n	k	0.1	0.2	0.3	0.4	0.5	k
2	0	0.8100	0.6400	0.4900	0.3600	0.2500	0
	1	0.9900	0.9600	0.9100	0.8400	0.7500	1
	2	1.0000	1.0000	1.0000	1.0000	1.0000	2
3	0	0.7290	0.5120	0.3430	0.2160	0.1250	0
	1	0.9720	0.8960	0.7840	0.6480	0.5000	1
	2	0.9990	0.9920	0.9730	0.9360	0.8750	2
	3	1.0000	1.0000	1.0000	1.0000	1.0000	3
4	0	0.6561	0.4096	0.2401	0.1296	0.0625	0
	1	0.9477	0.8192	0.6517	0.4752	0.3125	1
	2	0.9963	0.9728	0.9163	0.8208	0.6875	2
	3	0.9999	0.9984	0.9919	0.9744	0.9375	3
	4	1.0000	1.0000	1.0000	1.0000	1.0000	4
5	0	0.5905	0.3277	0.1681	0.0778	0.0313	0
	1	0.9185	0.7373	0.5282	0.3370	0.1875	1
	2	0.9914	0.9421	0.8369	0.6826	0.5000	2
	3	0.9995	0.9933	0.9692	0.9130	0.8125	3
	4	1.0000	0.9997	0.9976	0.9898	0.9687	4
	5	1.0000	1.0000	1.0000	1.0000	1.0000	5
6	0	0.5314	0.2621	0.1176	0.0467	0.0156	0
	1	0.8857	0.6554	0.4202	0.2333	0.1094	1
	2	0.9842	0.9011	0.7443	0.5443	0.3438	2
	3	0.9987	0.9830	0.9295	0.8208	0.6563	3
	4	0.9999	0.9984	0.9891	0.9590	0.8906	4
	5	1.0000	0.9999	0.9993	0.9959	0.9844	5
	6	1.0000	1.0000	1.0000	1.0000	1.0000	6
7	0	0.4783	0.2097	0.0824	0.0280	0.0078	0
	1	0.8503	0.5767	0.3294	0.1586	0.0625	1
	2	0.9743	0.8520	0.6471	0.4199	0.2266	2
	3	0.9973	0.9667	0.8740	0.7102	0.5000	3
	4	0.9998	0.9953	0.9712	0.9037	0.7734	4
	5	1.0000	0.9996	0.9962	0.9812	0.9375	5
	6	1.0000	1.0000	0.9998	0.9984	0.9922	6
	7	1.0000	1.0000	1.0000	1.0000	1.0000	7

Table 24 THE CUMULATIVE BINOMIAL DISTRIBUTION

(Description: p. 60)

n	k	binomial probability, p: 0.1	0.2	0.3	0.4	0.5	k
8	0	0.4305	0.1678	0.0576	0.0168	0.0039	0
	1	0.8131	0.5033	0.2553	0.1064	0.0352	1
	2	0.9619	0.7969	0.5518	0.3154	0.1445	2
	3	0.9950	0.9437	0.8059	0.5941	0.3633	3
	4	0.9996	0.9896	0.9420	0.8263	0.6367	4
	5	1.0000	0.9988	0.9887	0.9502	0.8555	5
	6	1.0000	0.9999	0.9987	0.9915	0.9648	6
	7	1.0000	1.0000	0.9999	0.9993	0.9961	7
	8	1.0000	1.0000	1.0000	1.0000	1.0000	8
9	0	0.3874	0.1342	0.0404	0.0101	0.0020	0
	1	0.7748	0.4362	0.1960	0.0705	0.0195	1
	2	0.9470	0.7382	0.4628	0.2318	0.0898	2
	3	0.9917	0.9144	0.7297	0.4826	0.2539	3
	4	0.9991	0.9804	0.9012	0.7334	0.5000	4
	5	0.9999	0.9969	0.9747	0.9006	0.7461	5
	6	1.0000	0.9997	0.9957	0.9750	0.9102	6
	7	1.0000	1.0000	0.9996	0.9962	0.9805	7
	8	1.0000	1.0000	1.0000	0.9997	0.9980	8
	9	1.0000	1.0000	1.0000	1.0000	1.0000	9
10	0	0.3487	0.1074	0.0282	0.0060	0.0010	0
	1	0.7361	0.3758	0.1493	0.0464	0.0107	1
	2	0.9298	0.6778	0.3828	0.1673	0.0547	2
	3	0.9872	0.8791	0.6496	0.3823	0.1719	3
	4	0.9984	0.9672	0.8497	0.6331	0.3770	4
	5	0.9999	0.9936	0.9527	0.8338	0.6230	5
	6	1.0000	0.9991	0.9894	0.9452	0.8281	6
	7	1.0000	0.9999	0.9984	0.9877	0.9453	7
	8	1.0000	1.0000	0.9999	0.9983	0.9893	8
	9	1.0000	1.0000	1.0000	0.9999	0.9990	9
	10	1.0000	1.0000	1.0000	1.0000	1.0000	10
11	0	0.3138	0.0859	0.0198	0.0036	0.0005	0
	1	0.6974	0.3221	0.1130	0.0302	0.0059	1
	2	0.9104	0.6174	0.3127	0.1189	0.0327	2
	3	0.9815	0.8389	0.5696	0.2963	0.1133	3
	4	0.9972	0.9496	0.7897	0.5328	0.2744	4
	5	0.9997	0.9883	0.9218	0.7535	0.5000	5
	6	1.0000	0.9980	0.9784	0.9006	0.7256	6
	7	1.0000	0.9998	0.9957	0.9707	0.8867	7
	8	1.0000	1.0000	0.9994	0.9941	0.9673	8
	9	1.0000	1.0000	1.0000	0.9993	0.9941	9
	10	1.0000	1.0000	1.0000	1.0000	0.9995	10
	11	1.0000	1.0000	1.0000	1.0000	1.0000	11

Table 24 THE CUMULATIVE BINOMIAL DISTRIBUTION
(Description: p. 60)

n	k	0.1	0.2	0.3	0.4	0.5	k
		binomial probability, p					
12	0	0.2824	0.0687	0.0138	0.0022	0.0002	0
	1	0.6590	0.2749	0.0850	0.0196	0.0032	1
	2	0.8891	0.5583	0.2528	0.0834	0.0193	2
	3	0.9744	0.7946	0.4925	0.2253	0.0730	3
	4	0.9957	0.9274	0.7237	0.4382	0.1938	4
	5	0.9995	0.9806	0.8822	0.6652	0.3872	5
	6	0.9999	0.9961	0.9614	0.8418	0.6128	6
	7	1.0000	0.9994	0.9905	0.9427	0.8062	7
	8	1.0000	0.9999	0.9983	0.9847	0.9270	8
	9	1.0000	1.0000	0.9998	0.9972	0.9807	9
	10	1.0000	1.0000	1.0000	0.9997	0.9968	10
	11	1.0000	1.0000	1.0000	1.0000	0.9998	11
	12	1.0000	1.0000	1.0000	1.0000	1.0000	12
13	0	0.2542	0.0550	0.0097	0.0013	0.0001	0
	1	0.6213	0.2336	0.0637	0.0126	0.0017	1
	2	0.8661	0.5017	0.2025	0.0579	0.0112	2
	3	0.9658	0.7473	0.4206	0.1686	0.0461	3
	4	0.9935	0.9009	0.6543	0.3530	0.1334	4
	5	0.9991	0.9700	0.8346	0.5744	0.2905	5
	6	0.9999	0.9930	0.9376	0.7712	0.5000	6
	7	1.0000	0.9988	0.9818	0.9023	0.7095	7
	8	1.0000	0.9998	0.9960	0.9679	0.8666	8
	9	1.0000	1.0000	0.9993	0.9922	0.9539	9
	10	1.0000	1.0000	0.9999	0.9987	0.9888	10
	11	1.0000	1.0000	1.0000	0.9999	0.9983	11
	12	1.0000	1.0000	1.0000	1.0000	0.9999	12
	13	1.0000	1.0000	1.0000	1.0000	1.0000	13
14	0	0.2288	0.0440	0.0068	0.0008	0.0001	0
	1	0.5846	0.1979	0.0475	0.0081	0.0009	1
	2	0.8416	0.4481	0.1608	0.0398	0.0065	2
	3	0.9559	0.6982	0.3552	0.1243	0.0287	3
	4	0.9908	0.8702	0.5842	0.2793	0.0898	4
	5	0.9985	0.9561	0.7805	0.4859	0.2120	5
	6	0.9998	0.9884	0.9067	0.6925	0.3953	6
	7	1.0000	0.9976	0.9685	0.8499	0.6047	7
	8	1.0000	0.9996	0.9917	0.9417	0.7880	8
	9	1.0000	1.0000	0.9983	0.9825	0.9102	9
	10	1.0000	1.0000	0.9998	0.9961	0.9713	10
	11	1.0000	1.0000	1.0000	0.9994	0.9935	11
	12	1.0000	1.0000	1.0000	0.9999	0.9991	12
	13	1.0000	1.0000	1.0000	1.0000	0.9999	13
	14	1.0000	1.0000	1.0000	1.0000	1.0000	14

Table 24 THE CUMULATIVE BINOMIAL DISTRIBUTION (Description: p. 60)

n	k	0.1	0.2	0.3	0.4	0.5	k
		binomial probability, p					
15	0	0.2059	0.0352	0.0047	0.0005	0.0000	0
	1	0.5490	0.1671	0.0353	0.0052	0.0005	1
	2	0.8159	0.3980	0.1268	0.0271	0.0037	2
	3	0.9444	0.6482	0.2969	0.0905	0.0176	3
	4	0.9873	0.8358	0.5155	0.2173	0.0592	4
	5	0.9978	0.9389	0.7216	0.4032	0.1509	5
	6	0.9997	0.9819	0.8689	0.6098	0.3036	6
	7	1.0000	0.9958	0.9500	0.7869	0.5000	7
	8	1.0000	0.9992	0.9848	0.9050	0.6964	8
	9	1.0000	0.9999	0.9963	0.9662	0.8491	9
	10	1.0000	1.0000	0.9993	0.9907	0.9408	10
	11	1.0000	1.0000	0.9999	0.9981	0.9824	11
	12	1.0000	1.0000	1.0000	0.9997	0.9963	12
	13	1.0000	1.0000	1.0000	1.0000	0.9995	13
	14	1.0000	1.0000	1.0000	1.0000	1.0000	14
	15	1.0000	1.0000	1.0000	1.0000	1.0000	15
16	0	0.1853	0.0281	0.0033	0.0003	0.0000	0
	1	0.5147	0.1407	0.0261	0.0033	0.0003	1
	2	0.7892	0.3518	0.0994	0.0183	0.0021	2
	3	0.9316	0.5981	0.2459	0.0651	0.0106	3
	4	0.9830	0.7982	0.4499	0.1666	0.0384	4
	5	0.9967	0.9183	0.6598	0.3288	0.1051	5
	6	0.9995	0.9733	0.8247	0.5272	0.2272	6
	7	0.9999	0.9930	0.9256	0.7161	0.4018	7
	8	1.0000	0.9985	0.9743	0.8577	0.5982	8
	9	1.0000	0.9998	0.9929	0.9417	0.7728	9
	10	1.0000	1.0000	0.9984	0.9809	0.8949	10
	11	1.0000	1.0000	0.9997	0.9951	0.9616	11
	12	1.0000	1.0000	1.0000	0.9991	0.9894	12
	13	1.0000	1.0000	1.0000	0.9999	0.9979	13
	14	1.0000	1.0000	1.0000	1.0000	0.9997	14
	15	1.0000	1.0000	1.0000	1.0000	1.0000	15
17	0	0.1668	0.0225	0.0023	0.0002	0.0000	0
	1	0.4818	0.1182	0.0193	0.0021	0.0001	1
	2	0.7618	0.3096	0.0774	0.0123	0.0012	2
	3	0.9174	0.5489	0.2019	0.0464	0.0064	3
	4	0.9779	0.7582	0.3887	0.1260	0.0245	4
	5	0.9953	0.8943	0.5968	0.2639	0.0717	5
	6	0.9992	0.9623	0.7752	0.4478	0.1662	6
	7	0.9999	0.9891	0.8954	0.6405	0.3145	7

Table 24 THE CUMULATIVE BINOMIAL DISTRIBUTION
(Description: p. 60)

		binomial probability, p					
n	k	0.1	0.2	0.3	0.4	0.5	k
17	8	1.0000	0.9974	0.9597	0.8011	0.5000	8
	9	1.0000	0.9995	0.9873	0.9081	0.6855	9
	10	1.0000	0.9999	0.9968	0.9652	0.8338	10
	11	1.0000	1.0000	0.9993	0.9894	0.9283	11
	12	1.0000	1.0000	0.9999	0.9975	0.9755	12
	13	1.0000	1.0000	1.0000	0.9995	0.9936	13
	14	1.0000	1.0000	1.0000	0.9999	0.9988	14
	15	1.0000	1.0000	1.0000	1.0000	0.9999	15
	16	1.0000	1.0000	1.0000	1.0000	1.0000	16
18	0	0.1501	0.0180	0.0016	0.0001	0.0000	0
	1	0.4503	0.0991	0.0142	0.0013	0.0001	1
	2	0.7338	0.2713	0.0600	0.0082	0.0007	2
	3	0.9018	0.5010	0.1646	0.0328	0.0038	3
	4	0.9718	0.7164	0.3327	0.0942	0.0154	4
	5	0.9936	0.8671	0.5344	0.2088	0.0481	5
	6	0.9988	0.9487	0.7217	0.3743	0.1189	6
	7	0.9998	0.9837	0.8593	0.5634	0.2403	7
	8	1.0000	0.9957	0.9404	0.7368	0.4073	8
	9	1.0000	0.9991	0.9790	0.8653	0.5927	9
	10	1.0000	0.9998	0.9939	0.9424	0.7597	10
	11	1.0000	1.0000	0.9986	0.9797	0.8811	11
	12	1.0000	1.0000	0.9997	0.9942	0.9519	12
	13	1.0000	1.0000	1.0000	0.9987	0.9846	13
	14	1.0000	1.0000	1.0000	0.9998	0.9962	14
	15	1.0000	1.0000	1.0000	1.0000	0.9993	15
	16	1.0000	1.0000	1.0000	1.0000	0.9999	16
	17	1.0000	1.0000	1.0000	1.0000	1.0000	17
19	0	0.1351	0.0144	0.0011	0.0001	0.0000	0
	1	0.4203	0.0829	0.0104	0.0008	0.0000	1
	2	0.7054	0.2369	0.0462	0.0055	0.0004	2
	3	0.8850	0.4551	0.1332	0.0230	0.0022	3
	4	0.9648	0.6733	0.2822	0.0696	0.0096	4
	5	0.9914	0.8369	0.4739	0.1629	0.0318	5
	6	0.9983	0.9324	0.6655	0.3081	0.0835	6
	7	0.9997	0.9767	0.8180	0.4878	0.1796	7
	8	1.0000	0.9933	0.9161	0.6675	0.3238	8
	9	1.0000	0.9984	0.9674	0.8139	0.5000	9
	10	1.0000	0.9997	0.9895	0.9115	0.6762	10
	11	1.0000	1.0000	0.9972	0.9648	0.8204	11
	12	1.0000	1.0000	0.9994	0.9884	0.9165	12

Table 24 THE CUMULATIVE BINOMIAL DISTRIBUTION
(Description: p. 60)

		binomial probability, p					
n	k	0.1	0.2	0.3	0.4	0.5	k
19	13	1.0000	1.0000	0.9999	0.9969	0.9682	13
	14	1.0000	1.0000	1.0000	0.9994	0.9904	14
	15	1.0000	1.0000	1.0000	0.9999	0.9978	15
	16	1.0000	1.0000	1.0000	1.0000	0.9996	16
	17	1.0000	1.0000	1.0000	1.0000	1.0000	17
20	0	0.1216	0.0115	0.0008	0.0000	0.0000	0
	1	0.3917	0.0692	0.0076	0.0005	0.0000	1
	2	0.6769	0.2061	0.0355	0.0036	0.0002	2
	3	0.8670	0.4114	0.1071	0.0160	0.0013	3
	4	0.9568	0.6296	0.2375	0.0510	0.0059	4
	5	0.9887	0.8042	0.4164	0.1256	0.0207	5
	6	0.9976	0.9133	0.6080	0.2500	0.0577	6
	7	0.9996	0.9679	0.7723	0.4159	0.1316	7
	8	0.9999	0.9900	0.8867	0.5956	0.2517	8
	9	1.0000	0.9974	0.9520	0.7553	0.4119	9
	10	1.0000	0.9994	0.9829	0.8725	0.5881	10
	11	1.0000	0.9999	0.9949	0.9435	0.7483	11
	12	1.0000	1.0000	0.9987	0.9790	0.8684	12
	13	1.0000	1.0000	0.9997	0.9935	0.9423	13
	14	1.0000	1.0000	1.0000	0.9984	0.9793	14
	15	1.0000	1.0000	1.0000	0.9997	0.9941	15
	16	1.0000	1.0000	1.0000	1.0000	0.9987	16
	17	1.0000	1.0000	1.0000	1.0000	0.9998	17
	18	1.0000	1.0000	1.0000	1.0000	1.0000	18
21	0	0.1094	0.0092	0.0006	0.0000	0.0000	0
	1	0.3647	0.0576	0.0056	0.0003	0.0000	1
	2	0.6484	0.1787	0.0271	0.0024	0.0001	2
	3	0.8480	0.3704	0.0856	0.0110	0.0007	3
	4	0.9478	0.5860	0.1984	0.0370	0.0036	4
	5	0.9856	0.7693	0.3627	0.0957	0.0133	5
	6	0.9967	0.8915	0.5505	0.2002	0.0392	6
	7	0.9994	0.9569	0.7230	0.3495	0.0946	7
	8	0.9999	0.9856	0.8523	0.5237	0.1917	8
	9	1.0000	0.9959	0.9324	0.6914	0.3318	9
	10	1.0000	0.9990	0.9736	0.8256	0.5000	10
	11	1.0000	0.9998	0.9913	0.9151	0.6682	11
	12	1.0000	1.0000	0.9976	0.9648	0.8083	12
	13	1.0000	1.0000	0.9994	0.9877	0.9054	13
	14	1.0000	1.0000	0.9999	0.9964	0.9608	14

Table 24 THE CUMULATIVE BINOMIAL DISTRIBUTION
(Description: p. 60)

		binomial probability, p					
n	k	0.1	0.2	0.3	0.4	0.5	k
21	15	1.0000	1.0000	1.0000	0.9992	0.9867	15
	16	1.0000	1.0000	1.0000	0.9998	0.9964	16
	17	1.0000	1.0000	1.0000	1.0000	0.9993	17
	18	1.0000	1.0000	1.0000	1.0000	0.9999	18
	19	1.0000	1.0000	1.0000	1.0000	1.0000	19
22	0	0.0985	0.0074	0.0004	0.0000	0.0000	0
	1	0.3392	0.0480	0.0041	0.0002	0.0000	1
	2	0.6200	0.1545	0.0207	0.0016	0.0001	2
	3	0.8281	0.3320	0.0681	0.0076	0.0004	3
	4	0.9379	0.5429	0.1645	0.0266	0.0022	4
	5	0.9818	0.7326	0.3134	0.0722	0.0085	5
	6	0.9956	0.8670	0.4942	0.1584	0.0262	6
	7	0.9991	0.9439	0.6713	0.2898	0.0669	7
	8	0.9999	0.9799	0.8135	0.4540	0.1431	8
	9	1.0000	0.9939	0.9084	0.6244	0.2617	9
	10	1.0000	0.9984	0.9613	0.7720	0.4159	10
	11	1.0000	0.9997	0.9860	0.8793	0.5841	11
	12	1.0000	0.9999	0.9957	0.9449	0.7383	12
	13	1.0000	1.0000	0.9989	0.9785	0.8569	13
	14	1.0000	1.0000	0.9998	0.9930	0.9331	14
	15	1.0000	1.0000	1.0000	0.9981	0.9738	15
	16	1.0000	1.0000	1.0000	0.9996	0.9915	16
	17	1.0000	1.0000	1.0000	0.9999	0.9978	17
	18	1.0000	1.0000	1.0000	1.0000	0.9996	18
	19	1.0000	1.0000	1.0000	1.0000	0.9999	19
	20	1.0000	1.0000	1.0000	1.0000	1.0000	20
23	0	0.0886	0.0059	0.0003	0.0000	0.0000	0
	1	0.3151	0.0398	0.0030	0.0001	0.0000	1
	2	0.5920	0.1332	0.0157	0.0010	0.0000	2
	3	0.8073	0.2965	0.0538	0.0052	0.0002	3
	4	0.9269	0.5007	0.1356	0.0190	0.0013	4
	5	0.9774	0.6947	0.2688	0.0540	0.0053	5
	6	0.9942	0.8402	0.4399	0.1240	0.0173	6
	7	0.9988	0.9285	0.6181	0.2373	0.0466	7
	8	0.9998	0.9727	0.7709	0.3884	0.1050	8
	9	1.0000	0.9911	0.8799	0.5562	0.2024	9
	10	1.0000	0.9975	0.9454	0.7129	0.3388	10
	11	1.0000	0.9994	0.9786	0.8364	0.5000	11
	12	1.0000	0.9999	0.9928	0.9187	0.6612	12
	13	1.0000	1.0000	0.9979	0.9651	0.7976	13
	14	1.0000	1.0000	0.9995	0.9872	0.8950	14

Table 24 THE CUMULATIVE BINOMIAL DISTRIBUTION
(Description: p. 60)

n	k	binomial probability, p 0.1	0.2	0.3	0.4	0.5	k
23	15	1.0000	1.0000	0.9999	0.9960	0.9534	15
	16	1.0000	1.0000	1.0000	0.9990	0.9827	16
	17	1.0000	1.0000	1.0000	0.9998	0.9947	17
	18	1.0000	1.0000	1.0000	1.0000	0.9987	18
	19	1.0000	1.0000	1.0000	1.0000	0.9998	19
	20	1.0000	1.0000	1.0000	1.0000	1.0000	20
24	0	0.0798	0.0047	0.0002	0.0000	0.0000	0
	1	0.2925	0.0331	0.0022	0.0001	0.0000	1
	2	0.5643	0.1145	0.0119	0.0007	0.0000	2
	3	0.7857	0.2639	0.0424	0.0035	0.0001	3
	4	0.9149	0.4599	0.1111	0.0134	0.0008	4
	5	0.9723	0.6559	0.2288	0.0400	0.0033	5
	6	0.9925	0.8111	0.3886	0.0960	0.0113	6
	7	0.9983	0.9108	0.5647	0.1919	0.0320	7
	8	0.9997	0.9638	0.7250	0.3279	0.0758	8
	9	0.9999	0.9874	0.8472	0.4891	0.1537	9
	10	1.0000	0.9962	0.9258	0.6502	0.2706	10
	11	1.0000	0.9990	0.9686	0.7870	0.4194	11
	12	1.0000	0.9998	0.9885	0.8857	0.5806	12
	13	1.0000	1.0000	0.9964	0.9465	0.7294	13
	14	1.0000	1.0000	0.9990	0.9783	0.8463	14
	15	1.0000	1.0000	0.9998	0.9925	0.9242	15
	16	1.0000	1.0000	1.0000	0.9978	0.9680	16
	17	1.0000	1.0000	1.0000	0.9995	0.9887	17
	18	1.0000	1.0000	1.0000	0.9999	0.9967	18
	19	1.0000	1.0000	1.0000	1.0000	0.9992	19
	20	1.0000	1.0000	1.0000	1.0000	0.9999	20
	21	1.0000	1.0000	1.0000	1.0000	1.0000	21
25	0	0.0718	0.0038	0.0001	0.0000	0.0000	0
	1	0.2712	0.0274	0.0016	0.0001	0.0000	1
	2	0.5371	0.0982	0.0090	0.0004	0.0000	2
	3	0.7636	0.2340	0.0332	0.0024	0.0001	3
	4	0.9020	0.4207	0.0905	0.0095	0.0005	4
	5	0.9666	0.6167	0.1935	0.0294	0.0020	5
	6	0.9905	0.7800	0.3407	0.0736	0.0073	6
	7	0.9977	0.8909	0.5118	0.1536	0.0216	7
	8	0.9995	0.9532	0.6769	0.2735	0.0539	8
	9	0.9999	0.9827	0.8106	0.4246	0.1148	9
	10	1.0000	0.9944	0.9022	0.5858	0.2122	10
	11	1.0000	0.9985	0.9558	0.7323	0.3450	11
	12	1.0000	0.9996	0.9825	0.8462	0.5000	12

Table 24 THE CUMULATIVE BINOMIAL DISTRIBUTION
(Description: p. 60)

		binomial probability, p					
n	k	0.1	0.2	0.3	0.4	0.5	k
25	13	1.0000	0.9999	0.9940	0.9222	0.6550	13
	14	1.0000	1.0000	0.9982	0.9656	0.7878	14
	15	1.0000	1.0000	0.9995	0.9868	0.8852	15
	16	1.0000	1.0000	0.9999	0.9957	0.9461	16
	17	1.0000	1.0000	1.0000	0.9988	0.9784	17
	18	1.0000	1.0000	1.0000	0.9997	0.9927	18
	19	1.0000	1.0000	1.0000	0.9999	0.9980	19
	20	1.0000	1.0000	1.0000	1.0000	0.9995	20
	21	1.0000	1.0000	1.0000	1.0000	0.9999	21
	22	1.0000	1.0000	1.0000	1.0000	1.0000	22
30	0	0.0424	0.0012	0.0000	0.0000	0.0000	0
	1	0.1837	0.0105	0.0003	0.0000	0.0000	1
	2	0.4114	0.0442	0.0021	0.0000	0.0000	2
	3	0.6474	0.1227	0.0093	0.0003	0.0000	3
	4	0.8245	0.2552	0.0302	0.0015	0.0000	4
	5	0.9268	0.4275	0.0766	0.0057	0.0002	5
	6	0.9742	0.6070	0.1595	0.0172	0.0007	6
	7	0.9922	0.7608	0.2814	0.0435	0.0026	7
	8	0.9980	0.8713	0.4315	0.0940	0.0081	8
	9	0.9995	0.9389	0.5888	0.1763	0.0214	9
	10	0.9999	0.9744	0.7304	0.2915	0.0494	10
	11	1.0000	0.9905	0.8407	0.4311	0.1002	11
	12	1.0000	0.9969	0.9155	0.5785	0.1808	12
	13	1.0000	0.9991	0.9599	0.7145	0.2923	13
	14	1.0000	0.9998	0.9831	0.8246	0.4278	14
	15	1.0000	0.9999	0.9936	0.9029	0.5722	15
	16	1.0000	1.0000	0.9979	0.9519	0.7077	16
	17	1.0000	1.0000	0.9994	0.9788	0.8192	17
	18	1.0000	1.0000	0.9998	0.9917	0.8998	18
	19	1.0000	1.0000	1.0000	0.9971	0.9506	19
	20	1.0000	1.0000	1.0000	0.9991	0.9786	20
	21	1.0000	1.0000	1.0000	0.9998	0.9919	21
	22	1.0000	1.0000	1.0000	1.0000	0.9974	22
	23	1.0000	1.0000	1.0000	1.0000	0.9993	23
	24	1.0000	1.0000	1.0000	1.0000	0.9998	24
	25	1.0000	1.0000	1.0000	1.0000	1.0000	25

Table 25 CONFIDENCE LIMITS FOR A BINOMIAL PARAMETER

(Table: pp. 71-79)

A binomial sample of size n results in k successes. A confidence interval for the true proportion of successes with confidence γ (of 0.90, 0.95 or 0.99) is desired. The tabulated entries have the upper confidence limit above the lower confidence limit. Example: Four successes in ten trials give a 95% confidence interval of (0.150, 0.696). Confidence limits for n - k successes in n trials may be found by first finding the confidence limit for k successes in n trials, say (P_L, P_u), and then forming the interval $(1 - P_u, 1 - P_L)$.

Table 25 CONFIDENCE LIMITS ON A BINOMIAL PARAMETER
(Description: p. 70)

$\gamma = .90$

k	sample size, n							
	2	3	4	5	6	7	8	9
0	0.776 0.000	0.632 0.000	0.527 0.000	0.451 0.000	0.393 0.000	0.348 0.000	0.312 0.000	0.283 0.000
1	0.975 0.025	0.865 0.017	0.751 0.013	0.657 0.010	0.582 0.009	0.521 0.007	0.471 0.006	0.429 0.006
2	1.000 0.224	0.983 0.135	0.902 0.098	0.811 0.076	0.729 0.063	0.659 0.053	0.600 0.046	0.550 0.041
3		1.000 0.368	0.987 0.249	0.924 0.189	0.847 0.153	0.775 0.129	0.711 0.111	0.655 0.098
4			1.000 0.473	0.990 0.343	0.937 0.271	0.871 0.225	0.807 0.193	0.749 0.169

k	sample size, n							
	10	11	12	13	14	15	16	17
0	0.259 0.000	0.238 0.000	0.221 0.000	0.206 0.000	0.193 0.000	0.181 0.000	0.171 0.000	0.162 0.000
1	0.394 0.005	0.364 0.005	0.339 0.004	0.316 0.004	0.297 0.004	0.279 0.003	0.264 0.003	0.250 0.003
2	0.507 0.037	0.470 0.033	0.438 0.030	0.410 0.028	0.385 0.026	0.363 0.024	0.344 0.023	0.326 0.021
3	0.607 0.087	0.564 0.079	0.527 0.072	0.495 0.066	0.466 0.061	0.440 0.057	0.417 0.053	0.396 0.050
4	0.696 0.150	0.650 0.135	0.609 0.123	0.573 0.113	0.540 0.104	0.511 0.097	0.484 0.090	0.461 0.085
5	0.778 0.222	0.729 0.200	0.685 0.181	0.645 0.166	0.610 0.153	0.577 0.142	0.548 0.132	0.522 0.124
6	0.850 0.304	0.800 0.271	0.755 0.245	0.713 0.224	0.675 0.206	0.640 0.191	0.609 0.178	0.580 0.166
7	0.913 0.393	0.865 0.350	0.819 0.315	0.776 0.287	0.736 0.264	0.700 0.244	0.667 0.227	0.636 0.212
8	0.963 0.493	0.921 0.436	0.877 0.391	0.834 0.355	0.794 0.325	0.756 0.300	0.721 0.279	0.689 0.260

Table 25 CONFIDENCE LIMITS ON A BINOMIAL PARAMETER
(Description: p. 10)

$\gamma = .90$

	sample size, n							
k	18	19	20	22	24	26	28	30
0	0.153	0.146	0.139	0.127	0.117	0.109	0.101	0.095
	0.000	0.000	0.000	0.000	0.000	0.000	0.000	0.000
1	0.238	0.226	0.216	0.198	0.183	0.170	0.159	0.149
	0.003	0.003	0.003	0.002	0.002	0.002	0.002	0.002
2	0.310	0.296	0.283	0.259	0.240	0.223	0.208	0.195
	0.020	0.019	0.018	0.016	0.015	0.014	0.013	0.012
3	0.377	0.359	0.344	0.316	0.292	0.272	0.254	0.239
	0.047	0.044	0.042	0.038	0.035	0.032	0.030	0.028
4	0.439	0.419	0.401	0.369	0.342	0.318	0.298	0.280
	0.080	0.075	0.071	0.065	0.059	0.054	0.050	0.047
5	0.498	0.476	0.456	0.420	0.389	0.363	0.339	0.319
	0.116	0.110	0.104	0.094	0.086	0.079	0.073	0.068
6	0.554	0.530	0.508	0.468	0.435	0.405	0.380	0.357
	0.156	0.147	0.140	0.126	0.115	0.106	0.098	0.091
7	0.608	0.582	0.558	0.515	0.479	0.447	0.419	0.394
	0.199	0.188	0.177	0.160	0.146	0.134	0.124	0.115
8	0.659	0.632	0.606	0.561	0.521	0.487	0.457	0.430
	0.244	0.230	0.217	0.196	0.178	0.163	0.151	0.140
9	0.709	0.680	0.653	0.605	0.563	0.526	0.494	0.465
	0.291	0.274	0.259	0.233	0.212	0.194	0.179	0.166
10	0.756	0.726	0.698	0.647	0.603	0.564	0.530	0.499
	0.341	0.320	0.302	0.271	0.246	0.226	0.208	0.193
11	0.801	0.770	0.741	0.689	0.642	0.602	0.565	0.533
	0.392	0.368	0.347	0.311	0.282	0.258	0.238	0.221
12	0.844	0.812	0.783	0.729	0.681	0.638	0.600	0.566
	0.446	0.418	0.394	0.353	0.319	0.292	0.269	0.250
13	0.884	0.853	0.823	0.767	0.718	0.673	0.634	0.598
	0.502	0.470	0.442	0.395	0.358	0.327	0.301	0.279
14	0.920	0.890	0.860	0.804	0.754	0.708	0.667	0.630
	0.561	0.524	0.492	0.439	0.397	0.362	0.333	0.308
15	0.953	0.925	0.896	0.840	0.788	0.742	0.699	0.661
	0.623	0.581	0.544	0.485	0.437	0.398	0.366	0.339

Table 25 CONFIDENCE LIMITS ON A BINOMIAL PARAMETER
(Description: p. 70)

$\gamma = .90$

k	35	40	45	50	60	70	80	100
	sample size, n							
0	0.082	0.072	0.064	0.058	0.049	0.042	0.037	0.030
	0.000	0.000	0.000	0.000	0.000	0.000	0.000	0.000
1	0.129	0.113	0.101	0.091	0.077	0.066	0.058	0.047
	0.001	0.001	0.001	0.001	0.001	0.001	0.001	0.001
2	0.169	0.149	0.133	0.121	0.101	0.087	0.077	0.062
	0.010	0.009	0.008	0.007	0.006	0.005	0.004	0.004
3	0.207	0.183	0.163	0.148	0.124	0.107	0.094	0.076
	0.024	0.021	0.018	0.017	0.014	0.012	0.010	0.008
4	0.243	0.214	0.192	0.174	0.146	0.126	0.111	0.089
	0.040	0.035	0.031	0.028	0.023	0.020	0.017	0.014
5	0.277	0.245	0.220	0.199	0.167	0.144	0.127	0.102
	0.058	0.051	0.045	0.040	0.033	0.029	0.025	0.020
6	0.311	0.275	0.246	0.223	0.188	0.162	0.143	0.115
	0.077	0.067	0.060	0.054	0.044	0.038	0.033	0.026
7	0.343	0.304	0.272	0.247	0.208	0.180	0.158	0.127
	0.098	0.085	0.075	0.068	0.056	0.048	0.042	0.033
8	0.375	0.332	0.298	0.270	0.228	0.197	0.173	0.140
	0.119	0.104	0.092	0.082	0.068	0.058	0.051	0.040
9	0.406	0.360	0.323	0.293	0.247	0.214	0.188	0.152
	0.141	0.123	0.109	0.097	0.081	0.069	0.060	0.048
10	0.436	0.387	0.348	0.316	0.266	0.230	0.203	0.164
	0.164	0.142	0.126	0.113	0.093	0.080	0.069	0.055
12	0.496	0.440	0.396	0.360	0.304	0.263	0.232	0.187
	0.211	0.183	0.162	0.145	0.120	0.102	0.089	0.071
14	0.553	0.492	0.443	0.403	0.340	0.295	0.260	0.210
	0.260	0.226	0.199	0.178	0.147	0.125	0.109	0.087
16	0.608	0.542	0.489	0.445	0.376	0.326	0.288	0.233
	0.312	0.269	0.237	0.212	0.175	0.149	0.130	0.103
18	0.662	0.591	0.533	0.486	0.412	0.357	0.315	0.255
	0.365	0.315	0.277	0.247	0.204	0.173	0.151	0.120
20	0.714	0.639	0.577	0.526	0.446	0.388	0.342	0.277
	0.419	0.361	0.317	0.283	0.233	0.198	0.172	0.137

Table 25 CONFIDENCE LIMITS ON A BINOMIAL PARAMETER
(Description: p. 70)

γ = .95

sample size, n

k	2	3	4	5	6	7	8	9
0	0.842	0.708	0.602	0.522	0.459	0.410	0.369	0.336
	0.000	0.000	0.000	0.000	0.000	0.000	0.000	0.000
1	0.987	0.906	0.806	0.716	0.641	0.579	0.527	0.482
	0.013	0.008	0.006	0.005	0.004	0.004	0.003	0.003
2	1.000	0.992	0.932	0.853	0.777	0.710	0.651	0.600
	0.158	0.094	0.068	0.053	0.043	0.037	0.032	0.028
3		1.000	0.994	0.947	0.882	0.816	0.755	0.701
		0.292	0.194	0.147	0.118	0.099	0.085	0.075
4			1.000	0.995	0.957	0.901	0.843	0.788
			0.398	0.284	0.223	0.184	0.157	0.137

sample size, n

k	10	11	12	13	14	15	16	17
0	0.308	0.285	0.265	0.247	0.232	0.218	0.206	0.195
	0.000	0.000	0.000	0.000	0.000	0.000	0.000	0.000
1	0.445	0.413	0.385	0.360	0.339	0.319	0.302	0.287
	0.003	0.002	0.002	0.002	0.002	0.002	0.002	0.001
2	0.556	0.518	0.484	0.454	0.428	0.405	0.383	0.364
	0.025	0.023	0.021	0.019	0.018	0.017	0.016	0.015
3	0.652	0.610	0.572	0.538	0.508	0.481	0.456	0.434
	0.067	0.060	0.055	0.050	0.047	0.043	0.040	0.038
4	0.738	0.692	0.651	0.614	0.581	0.551	0.524	0.499
	0.122	0.109	0.099	0.091	0.084	0.078	0.073	0.068
5	0.813	0.766	0.723	0.684	0.649	0.616	0.587	0.560
	0.187	0.167	0.152	0.139	0.128	0.118	0.110	0.103
6	0.878	0.833	0.789	0.749	0.711	0.677	0.646	0.617
	0.262	0.234	0.211	0.192	0.177	0.163	0.152	0.142
7	0.933	0.891	0.848	0.808	0.770	0.734	0.701	0.671
	0.348	0.308	0.277	0.251	0.230	0.213	0.198	0.184
8	0.975	0.940	0.901	0.861	0.823	0.787	0.753	0.722
	0.444	0.390	0.349	0.316	0.289	0.266	0.247	0.230

Table 25 CONFIDENCE LIMITS ON A BINOMIAL PARAMETER
(Description: p. 70)

(Description: p. 70)

$\gamma = .95$

sample size, n

k	18	19	20	22	24	26	28	30
0	0.185 0.000	0.176 0.000	0.168 0.000	0.154 0.000	0.142 0.000	0.132 0.000	0.123 0.000	0.116 0.000
1	0.273 0.001	0.260 0.001	0.249 0.001	0.228 0.001	0.211 0.001	0.196 0.001	0.183 0.001	0.172 0.001
2	0.347 0.014	0.331 0.013	0.317 0.012	0.292 0.011	0.270 0.010	0.251 0.009	0.235 0.009	0.221 0.008
3	0.414 0.036	0.396 0.034	0.379 0.032	0.349 0.029	0.324 0.027	0.302 0.024	0.282 0.023	0.265 0.021
4	0.476 0.064	0.456 0.061	0.437 0.057	0.403 0.052	0.374 0.047	0.349 0.044	0.327 0.040	0.307 0.038
5	0.535 0.097	0.512 0.091	0.491 0.087	0.454 0.078	0.422 0.071	0.394 0.066	0.369 0.061	0.347 0.056
6	0.590 0.133	0.566 0.126	0.543 0.119	0.502 0.107	0.467 0.098	0.436 0.090	0.410 0.083	0.386 0.077
7	0.643 0.173	0.616 0.163	0.592 0.154	0.549 0.139	0.511 0.126	0.478 0.116	0.449 0.107	0.423 0.099
8	0.692 0.215	0.665 0.203	0.639 0.191	0.593 0.172	0.553 0.156	0.518 0.143	0.487 0.132	0.459 0.123
9	0.740 0.260	0.711 0.244	0.685 0.231	0.636 0.207	0.594 0.188	0.557 0.172	0.524 0.159	0.494 0.147
10	0.785 0.308	0.756 0.289	0.728 0.272	0.678 0.244	0.634 0.221	0.594 0.202	0.559 0.186	0.528 0.173
11	0.827 0.357	0.797 0.335	0.769 0.315	0.718 0.282	0.672 0.256	0.631 0.234	0.594 0.215	0.561 0.199
12	0.867 0.410	0.837 0.384	0.809 0.361	0.756 0.322	0.709 0.291	0.666 0.266	0.628 0.245	0.594 0.227
13	0.903 0.465	0.874 0.434	0.846 0.408	0.793 0.364	0.744 0.328	0.701 0.299	0.661 0.275	0.626 0.255
14	0.936 0.524	0.909 0.488	0.881 0.457	0.828 0.407	0.779 0.366	0.734 0.334	0.694 0.306	0.657 0.283
15	0.964 0.586	0.939 0.544	0.913 0.509	0.861 0.451	0.812 0.406	0.766 0.369	0.725 0.339	0.687 0.313

Table 25 CONFIDENCE LIMITS ON A BINOMIAL PARAMETER
(Description: p. 70)

$\gamma = .95$

sample size, n

k	35	40	45	50	60	70	80	100
0	0.100	0.088	0.079	0.071	0.060	0.051	0.045	0.036
	0.000	0.000	0.000	0.000	0.000	0.000	0.000	0.000
1	0.149	0.132	0.118	0.106	0.089	0.077	0.068	0.054
	0.001	0.001	0.001	0.001	0.000	0.000	0.000	0.000
2	0.192	0.169	0.151	0.137	0.115	0.099	0.087	0.070
	0.007	0.006	0.005	0.005	0.004	0.003	0.003	0.002
3	0.231	0.204	0.183	0.165	0.139	0.120	0.106	0.085
	0.018	0.016	0.014	0.013	0.010	0.009	0.008	0.006
4	0.267	0.237	0.212	0.192	0.162	0.140	0.123	0.099
	0.032	0.028	0.025	0.022	0.018	0.016	0.014	0.011
5	0.303	0.268	0.241	0.218	0.184	0.159	0.140	0.113
	0.048	0.042	0.037	0.033	0.028	0.024	0.021	0.016
6	0.336	0.298	0.268	0.243	0.205	0.177	0.156	0.126
	0.066	0.057	0.051	0.045	0.038	0.032	0.028	0.022
7	0.369	0.328	0.295	0.267	0.226	0.195	0.172	0.139
	0.084	0.073	0.065	0.058	0.048	0.041	0.036	0.029
8	0.401	0.356	0.321	0.291	0.246	0.213	0.188	0.152
	0.104	0.091	0.080	0.072	0.059	0.051	0.044	0.035
9	0.433	0.385	0.346	0.314	0.266	0.230	0.203	0.164
	0.125	0.108	0.096	0.086	0.071	0.061	0.053	0.042
10	0.463	0.412	0.371	0.337	0.285	0.247	0.218	0.176
	0.146	0.127	0.112	0.100	0.083	0.071	0.062	0.049
12	0.522	0.465	0.419	0.382	0.323	0.280	0.247	0.200
	0.191	0.166	0.146	0.131	0.108	0.092	0.080	0.064
14	0.579	0.517	0.466	0.425	0.360	0.313	0.276	0.224
	0.239	0.206	0.182	0.162	0.134	0.114	0.099	0.079
16	0.634	0.567	0.512	0.467	0.397	0.344	0.304	0.247
	0.288	0.249	0.219	0.195	0.161	0.137	0.119	0.094
18	0.686	0.615	0.557	0.508	0.432	0.376	0.332	0.269
	0.340	0.293	0.257	0.229	0.188	0.160	0.139	0.110
20	0.737	0.662	0.600	0.548	0.467	0.406	0.359	0.292
	0.394	0.338	0.296	0.264	0.217	0.184	0.160	0.127

Table 25 CONFIDENCE LIMITS ON A BINOMIAL PARAMETER
(Description: p. 70)

$\gamma = .99$

sample size, n

k	2	3	4	5	6	7	8	9
0	0.929 0.000	0.829 0.000	0.734 0.000	0.653 0.000	0.586 0.000	0.531 0.000	0.484 0.000	0.445 0.000
1	0.997 0.003	0.959 0.002	0.889 0.001	0.815 0.001	0.746 0.001	0.685 0.001	0.630 0.001	0.585 0.001
2	1.000 0.071	0.998 0.041	0.971 0.029	0.917 0.023	0.856 0.019	0.797 0.016	0.742 0.014	0.693 0.012
3		1.000 0.171	0.999 0.111	0.977 0.083	0.934 0.066	0.882 0.055	0.830 0.047	0.781 0.042
4			1.000 0.266	0.999 0.185	0.981 0.144	0.945 0.118	0.900 0.100	0.854 0.087

sample size, n

k	10	11	12	13	14	15	16	17
0	0.411 0.000	0.382 0.000	0.357 0.000	0.335 0.000	0.315 0.000	0.298 0.000	0.282 0.000	0.268 0.000
1	0.544 0.001	0.509 0.000	0.477 0.000	0.449 0.000	0.424 0.000	0.402 0.000	0.381 0.000	0.363 0.000
2	0.648 0.011	0.608 0.010	0.573 0.009	0.541 0.008	0.512 0.008	0.486 0.007	0.463 0.007	0.441 0.006
3	0.735 0.037	0.693 0.033	0.655 0.030	0.621 0.028	0.589 0.026	0.561 0.024	0.534 0.022	0.510 0.021
4	0.809 0.077	0.767 0.069	0.728 0.062	0.691 0.057	0.658 0.053	0.627 0.049	0.599 0.045	0.573 0.043
5	0.872 0.128	0.831 0.114	0.791 0.103	0.755 0.094	0.720 0.087	0.688 0.080	0.658 0.075	0.631 0.070
6	0.923 0.191	0.886 0.169	0.848 0.152	0.811 0.138	0.777 0.127	0.744 0.117	0.713 0.109	0.685 0.101
7	0.963 0.265	0.931 0.233	0.897 0.209	0.862 0.189	0.828 0.172	0.795 0.159	0.764 0.147	0.734 0.137
8	0.989 0.352	0.967 0.307	0.938 0.272	0.906 0.245	0.873 0.223	0.841 0.205	0.810 0.190	0.781 0.176

Table [illegible] CONFIDENCE LIMITS ON A BINOMIAL PARAMETER
(Description: p. 70)

γ = .99

sample size, n

k	18	19	20	22	24	26	28	30
0	0.255	0.243	0.233	0.214	0.198	0.184	0.172	0.162
	0.000	0.000	0.000	0.000	0.000	0.000	0.000	0.000
1	0.346	0.331	0.317	0.292	0.271	0.253	0.237	0.223
	0.000	0.000	0.000	0.000	0.000	0.000	0.000	0.000
2	0.422	0.404	0.387	0.358	0.332	0.310	0.291	0.274
	0.006	0.006	0.005	0.005	0.004	0.004	0.004	0.004
3	0.488	0.468	0.449	0.416	0.387	0.362	0.340	0.320
	0.020	0.019	0.018	0.016	0.015	0.013	0.012	0.012
4	0.549	0.527	0.507	0.470	0.438	0.410	0.385	0.363
	0.040	0.038	0.036	0.032	0.029	0.027	0.025	0.023
5	0.605	0.582	0.560	0.520	0.485	0.455	0.428	0.404
	0.065	0.062	0.058	0.053	0.048	0.044	0.041	0.038
6	0.658	0.633	0.610	0.567	0.530	0.498	0.469	0.443
	0.095	0.090	0.085	0.076	0.069	0.064	0.059	0.054
7	0.707	0.681	0.657	0.612	0.573	0.538	0.508	0.480
	0.128	0.121	0.114	0.102	0.093	0.085	0.079	0.073
8	0.753	0.726	0.701	0.655	0.614	0.578	0.545	0.516
	0.165	0.155	0.146	0.131	0.119	0.109	0.100	0.093
9	0.795	0.768	0.743	0.695	0.653	0.615	0.581	0.550
	0.205	0.192	0.181	0.162	0.146	0.134	0.123	0.114
10	0.835	0.808	0.782	0.734	0.690	0.651	0.616	0.583
	0.247	0.232	0.218	0.195	0.176	0.161	0.148	0.137
11	0.872	0.845	0.819	0.771	0.726	0.686	0.649	0.616
	0.293	0.274	0.257	0.229	0.207	0.189	0.173	0.160
12	0.905	0.879	0.854	0.805	0.760	0.719	0.681	0.647
	0.342	0.319	0.299	0.266	0.240	0.218	0.200	0.185
13	0.935	0.910	0.886	0.838	0.793	0.751	0.713	0.677
	0.395	0.367	0.343	0.305	0.274	0.249	0.228	0.211
14	0.960	0.938	0.915	0.869	0.824	0.782	0.743	0.707
	0.451	0.418	0.390	0.345	0.310	0.281	0.257	0.237
15	0.980	0.962	0.942	0.898	0.854	0.811	0.772	0.735
	0.512	0.473	0.440	0.388	0.347	0.314	0.287	0.265

Table 25 CONFIDENCE LIMITS ON A BINOMIAL PARAMETER
(Description: p. 70)

$\gamma = .99$

sample size, n

k	35	40	45	50	60	70	80	100
0	0.140	0.124	0.111	0.101	0.085	0.073	0.064	0.052
	0.000	0.000	0.000	0.000	0.000	0.000	0.000	0.000
1	0.194	0.172	0.154	0.139	0.117	0.101	0.089	0.072
	0.000	0.000	0.000	0.000	0.000	0.000	0.000	0.000
2	0.239	0.212	0.190	0.173	0.145	0.126	0.111	0.089
	0.003	0.003	0.002	0.002	0.002	0.001	0.001	0.001
3	0.280	0.248	0.223	0.203	0.171	0.148	0.131	0.105
	0.010	0.009	0.008	0.007	0.006	0.005	0.004	0.003
4	0.318	0.283	0.254	0.231	0.195	0.169	0.149	0.121
	0.020	0.017	0.015	0.014	0.011	0.010	0.009	0.007
5	0.354	0.315	0.284	0.258	0.218	0.189	0.167	0.135
	0.032	0.028	0.025	0.022	0.018	0.016	0.014	0.011
6	0.389	0.346	0.312	0.284	0.241	0.209	0.184	0.149
	0.046	0.040	0.036	0.032	0.026	0.023	0.020	0.016
7	0.422	0.376	0.339	0.309	0.262	0.227	0.201	0.163
	0.062	0.054	0.047	0.042	0.035	0.030	0.026	0.021
8	0.454	0.405	0.366	0.333	0.283	0.246	0.217	0.176
	0.079	0.068	0.060	0.054	0.045	0.038	0.033	0.026
9	0.485	0.434	0.392	0.357	0.304	0.264	0.233	0.189
	0.097	0.084	0.074	0.066	0.054	0.046	0.040	0.032
10	0.515	0.461	0.417	0.380	0.324	0.281	0.249	0.202
	0.115	0.100	0.088	0.079	0.065	0.055	0.048	0.038
12	0.574	0.514	0.466	0.425	0.362	0.315	0.279	0.227
	0.156	0.134	0.118	0.106	0.087	0.074	0.064	0.051
14	0.629	0.565	0.513	0.469	0.400	0.349	0.309	0.251
	0.199	0.171	0.151	0.134	0.110	0.094	0.081	0.065
16	0.681	0.614	0.558	0.511	0.437	0.381	0.338	0.275
	0.245	0.210	0.185	0.164	0.135	0.114	0.099	0.079
18	0.731	0.661	0.601	0.551	0.472	0.412	0.366	0.298
	0.294	0.252	0.220	0.196	0.161	0.136	0.118	0.093
20	0.778	0.705	0.643	0.591	0.507	0.443	0.394	0.321
	0.345	0.295	0.257	0.229	0.187	0.158	0.137	0.108

Table 26 SAMPLE SIZES FOR ONE-SIDED NONPARAMETRIC TOLERANCE LIMITS (Table: p. 81)

For any population the table gives the sample size n which will guarantee with confidence γ that the interval $(-\infty$, maximum observation$]$ contains a proportion p of the population. A similar statement holds for intervals of the form $[$minimum observation, $+\infty)$. The quantity tabulated is that value of n such that

$$p^n \leq 1 - \gamma$$

Table 26 SAMPLE SIZES FOR ONE-SIDED NONPARAMETRIC TOLERANCE LIMITS (Description: p. 80)

γ	p = 0.500	0.700	0.750	0.800	0.850	0.900	0.950
0.5000	1	2	3	4	5	7	14
0.7000	2	4	5	6	8	12	24
0.7500	2	4	5	7	9	14	28
0.8000	3	5	6	8	10	16	32
0.8500	3	6	7	9	12	19	37
0.9000	4	7	9	11	15	22	45
0.9500	5	9	11	14	19	29	59
0.9750	6	11	13	17	23	36	72
0.9800	6	11	14	18	25	38	77
0.9900	7	13	17	21	29	44	90
0.9950	8	15	19	24	33	51	104
0.9990	10	20	25	31	43	66	135
0.9995	11	22	27	35	47	73	149
0.9999	14	26	33	42	57	88	180

γ	p = 0.975	0.980	0.990	0.995	0.999	0.9995	0.9999
0.5000	28	35	69	139	693	1386	6932
0.7000	48	60	120	241	1204	2408	12040
0.7500	55	69	138	277	1386	2772	13863
0.8000	64	80	161	322	1609	3219	16094
0.8500	75	94	189	379	1897	3794	18971
0.9000	91	114	230	460	2302	4605	23025
0.9500	119	149	299	598	2995	5990	29956
0.9750	146	183	368	736	3688	7376	36887
0.9800	155	194	390	781	3911	7823	39119
0.9900	182	228	459	919	4603	9200	46050
0.9950	210	263	528	1058	5296	10594	52981
0.9990	273	342	688	1379	6905	13813	69075
0.9995	301	377	757	1517	7598	15199	76006
0.9999	364	456	917	1838	9206	18417	92099

Table 27 SAMPLE SIZES FOR TWO-SIDED NONPARAMETRIC TOLERANCE LIMITS (Table: p. 83)

For any population the table gives the sample size n which will guarantee with confidence γ that the interval [minimum observation, maximum observation] contains a proportion p of the population. The quantity tabulated is that value of n which satisfies

$$np^{n-1} - (n - 1)p^n \leq 1 - \gamma$$

Table 28 CRITICAL VALUES OF THE MANN-WHITNEY TWO-SAMPLE STATISTIC (Table: pp. 84-87)

Let X and Y be independent random variables with continuous distribution functions F(x), G(y), and let X_1, X_2, ..., X_n and Y_1, Y_2, ..., Y_m be random samples of X and Y. The Mann-Whitney statistic is defined as

$$U = \text{number of pairs } (X_i, Y_k) \text{ such that } X_i > Y_k$$

Under the hypothesis H: F(s) = G(s) for all s, the probability distribution of U does not depend on F(•). Table 28 contains for each three values (m, n, γ), the smallest integer u such that, under the hypothesis H, one has $P\{U \geq u\} \leq 1 - \gamma$. The same value u holds for the triple (n, m, γ). The one-sided critical values of u may be used to test H against the alternative that X is stochastically greater than Y (i.e., $F(s) \leq G(s)$ for all s and $F(s) < G(s)$ for some s). Wilcoxon used the statistic W, which is the sum of the ranks of the Y observations in the combined sample. The statistics U and W are related by $U = mn + \{m(m + 1)/2\} - W$. For a two-sided test, lower critical values can be obtained from the relationship $u^* = mn - u$, where

$$P\{U \leq u^*\} = P\{U \geq u\} \leq 1 - \gamma$$

Table 27 SAMPLE SIZES FOR TWO-SIDED NONPARAMETRIC TOLERANCE LIMITS (Description: p. 82)

				p			
γ	0.500	0.700	0.750	0.800	0.850	0.900	0.950
0.5000	3	6	7	9	11	17	34
0.7000	5	8	10	12	16	24	49
0.7500	5	9	10	13	18	27	53
0.8000	5	9	11	14	19	29	59
0.8500	6	10	13	16	22	33	67
0.9000	7	12	15	18	25	38	77
0.9500	8	14	18	22	30	46	93
0.9750	9	17	20	26	35	54	110
0.9800	9	17	21	27	37	56	115
0.9900	11	20	24	31	42	64	130
0.9950	12	22	27	34	47	72	146
0.9990	14	27	33	42	58	89	181
0.9995	15	29	36	46	63	96	196
0.9999	18	34	42	54	73	113	230

				p			
γ	0.975	0.980	0.990	0.995	0.999	0.9995	0.9999
0.5000	67	84	168	336	1679	3357	16784
0.7000	97	122	244	488	2439	4878	24392
0.7500	107	134	269	538	2692	5385	26926
0.8000	119	149	299	598	2994	5988	29943
0.8500	134	168	337	674	3372	6744	33724
0.9000	155	194	388	777	3889	7778	38896
0.9500	188	236	473	947	4742	9486	47437
0.9750	221	277	555	1113	5570	11142	55715
0.9800	231	290	581	1165	5832	11666	58337
0.9900	263	330	662	1325	6636	13274	66381
0.9950	294	369	740	1483	7427	14858	74299
0.9990	366	458	920	1843	9230	18463	92331
0.9995	396	496	996	1996	9995	19993	99983
0.9999	465	583	1171	2346	11751	23508	117559

Table 28 CRITICAL VALUES OF THE MANN WHITNEY TWO-SAMPLE STATISTIC (Description: p. 82)

		γ									
m	n	.70	.75	.80	.85	.90	.95	.975	.99	.995	.999
4	1	4	4	4	-	-	-	-	-	-	-
	2	6	7	7	7	8	-	-	-	-	-
	3	9	9	9	10	11	12	-	-	-	-
	4	11	11	12	13	13	15	16	-	-	-
5	1	5	5	5	-	-	-	-	-	-	-
5	2	7	8	8	9	9	10	-	-	-	-
	3	10	11	11	12	13	14	15	-	-	-
	4	13	14	15	15	16	18	19	20	-	-
	5	16	17	18	19	20	21	23	24	25	-
6	2	9	9	10	10	11	12	-	-	-	-
6	3	12	13	13	14	15	16	17	-	-	-
	4	16	16	17	18	19	21	22	23	24	-
	5	19	20	21	22	23	25	27	28	29	-
	6	22	23	24	26	27	29	31	33	34	-
7	2	10	10	11	12	13	14	-	-	-	-
7	3	14	15	15	16	17	19	20	21	-	-
	4	18	19	20	21	22	24	25	27	28	-
	5	22	23	24	25	27	29	30	32	34	-
	6	26	27	28	29	31	34	36	38	39	42
	7	30	31	32	34	36	38	41	43	45	48
8	3	16	16	17	18	19	21	22	24	-	-
	4	20	21	22	23	25	27	28	30	31	-
	5	25	26	27	28	30	32	34	36	38	40
	6	29	30	32	33	35	38	40	42	44	47
	7	34	35	36	38	40	43	46	49	50	54
8	8	38	40	41	43	45	49	51	55	57	60
9	3	17	18	19	20	22	23	25	26	27	-
	4	23	24	25	26	27	30	32	33	35	-
	5	28	29	30	31	33	36	38	40	42	44
	6	33	34	35	37	39	42	44	47	49	52
9	7	38	39	41	42	45	48	51	54	56	60
	8	43	44	46	48	50	54	57	61	63	67
	9	48	49	51	53	56	60	64	67	70	74
10	4	25	26	27	29	30	33	35	37	38	40
	5	30	32	33	35	37	39	42	44	46	49

Table 28 CRITICAL VALUES OF THE MANN-WHITNEY TWO-SAMPLE STATISTIC (Description: p. 82)

m	n	γ .70	.75	.80	.85	.90	.95	.975	.99	.995	.999
10	6	36	37	39	41	43	46	49	52	54	57
	7	42	43	45	47	49	53	56	59	61	65
	8	47	49	51	53	56	60	63	67	69	74
	9	53	54	57	59	62	66	70	74	77	82
	10	58	60	62	65	68	73	77	81	84	90
11	4	27	28	30	31	33	36	38	40	42	44
	5	33	35	36	38	40	43	46	48	50	53
	6	39	41	43	45	47	50	53	57	59	62
	7	45	47	49	51	54	58	61	65	67	71
	8	51	53	55	58	61	65	69	73	75	80
11	9	58	60	62	64	68	72	76	81	83	89
	10	64	66	68	71	74	79	84	88	92	98
	11	70	72	75	77	81	87	91	96	100	106
12	5	36	38	39	41	43	47	49	52	54	58
	6	43	44	46	48	51	55	58	61	63	68
12	7	49	51	53	55	58	63	66	70	72	77
	8	56	58	60	63	66	70	74	79	81	87
	9	63	65	67	70	73	78	82	87	90	96
	10	69	71	74	77	81	86	91	96	99	106
	11	76	78	81	84	88	94	99	104	108	115
12	12	82	85	88	91	95	102	107	113	117	124
13	5	39	41	42	44	47	50	53	56	58	62
	6	46	48	50	52	55	59	62	66	68	73
	7	53	55	57	60	63	67	71	75	78	83
	8	60	63	65	68	71	76	80	84	87	93
13	9	68	70	72	75	79	84	89	94	97	103
	10	75	77	80	83	87	93	97	103	106	113
	11	82	84	87	91	95	101	106	112	116	123
	12	89	92	95	98	103	109	115	121	125	133
	13	96	99	102	106	111	118	124	130	135	143
14	6	50	51	53	56	59	63	67	71	73	78
	7	57	59	62	64	67	72	76	81	83	89
	8	65	67	70	72	76	81	86	90	94	100
	9	72	75	78	81	85	90	95	100	104	111
	10	80	83	86	89	93	99	104	110	114	121

Table 28 CRITICAL VALUES OF THE MANN-WHITNEY TWO-SAMPLE STATISTIC (Description: p. 42)

m	n	γ .70	.75	.80	.85	.90	.95	.975	.99	.995	.999
14	11	88	91	94	97	102	108	114	120	124	132
	12	95	98	102	105	110	117	123	130	134	143
	13	103	106	110	114	119	126	132	139	144	153
	14	111	114	118	122	127	135	141	149	154	164
15	6	53	55	57	60	63	67	71	75	78	83
15	7	61	63	66	68	72	77	81	86	89	95
	8	69	72	74	77	81	87	91	96	100	106
	9	77	80	83	86	90	96	101	107	111	118
	10	86	88	91	95	99	106	111	117	121	129
	11	94	97	100	104	108	115	121	128	132	141
15	12	102	105	108	112	117	125	131	138	143	152
	13	110	113	117	121	127	134	141	148	153	163
	14	118	122	126	130	136	144	151	159	164	174
	15	126	130	134	139	145	153	161	169	174	185
16	7	65	67	70	73	76	82	86	91	94	101
16	8	74	76	79	82	86	92	97	102	106	113
	9	82	85	88	92	96	102	107	113	117	125
	10	91	94	97	101	106	112	118	124	129	137
	11	100	103	106	110	115	122	129	135	140	149
	12	108	112	115	120	125	132	139	146	151	161
16	13	117	121	124	129	134	143	149	157	163	173
	14	126	129	133	138	144	153	160	168	174	185
	15	134	138	143	147	154	163	170	179	185	197
	16	143	147	152	157	163	173	181	190	196	208
17	7	69	71	74	77	81	86	91	96	100	106
17	8	78	81	84	87	91	97	102	108	112	119
	9	87	90	93	97	101	108	114	120	124	132
	10	97	100	103	107	112	119	125	132	136	145
	11	106	109	113	117	122	130	136	143	148	158
	12	115	118	122	127	132	140	147	155	160	170
17	13	124	128	132	137	142	151	158	166	172	183
	14	133	137	141	146	153	161	169	178	184	195
	15	143	147	151	156	163	172	180	189	195	208
	16	152	156	161	166	173	183	191	201	207	220
	17	161	165	170	176	183	193	202	212	219	232

Table 28 CRITICAL VALUES OF THE MANN-WHITNEY TWO-SAMPLE STATISTIC (Description: p. 82)

m	n	γ .70	.75	.80	.85	.90	.95	.975	.99	.995	.999
18	8	83	85	88	92	96	103	108	114	118	126
	9	92	95	99	102	107	114	120	126	131	139
	10	102	105	109	113	118	125	132	139	143	153
	11	112	115	119	123	129	137	143	151	156	166
	12	122	125	129	134	139	148	155	163	169	179
18	13	131	135	139	144	150	159	167	175	181	192
	14	141	145	149	155	161	170	178	187	194	206
	15	151	155	160	165	172	182	190	200	206	219
	16	160	165	170	175	182	193	202	212	218	232
	17	170	175	180	186	193	204	213	224	231	245
18	18	180	185	190	196	204	215	225	236	243	258
19	8	87	90	93	97	101	108	114	120	124	132
	9	97	100	104	108	113	120	126	133	138	146
	10	108	111	115	119	124	132	138	146	151	161
	11	118	121	125	130	136	144	151	159	164	175
19	12	128	132	136	141	147	156	163	172	177	188
	13	138	142	147	152	158	167	175	184	190	202
	14	149	153	157	163	169	179	188	197	203	216
	15	159	163	168	174	181	191	200	210	216	230
	16	169	174	179	185	192	203	212	222	230	244
19	17	179	184	189	195	203	214	224	235	242	257
	18	189	194	200	206	214	226	236	248	255	271
	19	200	205	211	217	226	238	248	260	268	284
20	9	102	106	109	113	118	126	132	140	144	154
	10	113	117	120	125	130	138	145	153	158	168
20	11	124	128	132	136	142	151	158	167	172	183
	12	135	139	143	148	154	163	171	180	186	198
	13	145	150	154	159	166	176	184	193	200	212
	14	156	161	165	171	178	188	197	207	213	226
	15	167	171	177	182	190	200	210	220	227	241
20	16	178	182	188	194	201	213	222	233	241	255
	17	188	193	199	205	213	225	235	247	254	270
	18	199	204	210	217	225	237	248	260	268	284
	19	210	215	221	228	237	250	261	273	281	298
	20	221	226	232	240	249	262	273	286	295	312

Table 29 CRITICAL VALUES OF THE SIGN TEST AND DISTRIBUTION-FREE CONFIDENCE LIMITS ON THE MEDIAN (Table: p. 89)

The tabulated value is the largest m such that

$$\sum_{j=0}^{m} \binom{n}{j} \left(\frac{1}{2}\right)^{n} \leq \gamma$$

This table may be used for one- and two-sided sign test critical values as well as for setting distribution-free confidence limits for the median of a population. To construct a confidence interval for the median with confidence coefficient of at least $1 - \alpha$ find the table entry, say k, corresponding to the column $\gamma = \alpha/2$ and the sample size n. The interval runs between the $(k + 1)$st and $(n - k)$th order statistics of the sample.

Table 29 CRITICAL VALUES OF THE SIGN TEST AND DISTRIBUTION-FREE CONFIDENCE LIMITS ON THE MEDIAN (Description: p. 88)

	γ					
n	0.005	0.01	0.025	0.05	0.10	0.20
8	0	0	0	1	1	2
9	0	0	1	1	2	2
10	0	0	1	1	2	3
11	0	1	1	2	2	3
12	1	1	2	2	3	4
13	1	1	2	3	3	4
14	1	2	2	3	4	4
15	2	2	3	3	4	5
16	2	2	3	4	4	5
17	2	3	4	4	5	6
18	3	3	4	5	5	6
19	3	4	4	5	6	7
20	3	4	5	5	6	7
21	4	4	5	6	7	8
22	4	5	5	6	7	8
23	4	5	6	7	7	8
24	5	5	6	7	8	9
25	5	6	7	7	8	9
26	6	6	7	8	9	10
27	6	7	7	8	9	10
28	6	7	8	9	10	11
29	7	7	8	9	10	11
30	7	8	9	10	10	12
31	7	8	9	10	11	12
32	8	8	9	10	11	13
33	8	9	10	11	12	13
34	9	9	10	11	12	14
35	9	10	11	12	13	14
36	9	10	11	12	13	14
37	10	10	12	13	14	15
38	10	11	12	13	14	15
39	11	11	12	13	15	16
40	11	12	13	14	15	16
41	11	12	13	14	15	17
42	12	13	14	15	16	17
43	12	13	14	15	16	18
44	13	13	15	16	17	18
45	13	14	15	16	17	19
46	13	14	15	16	18	19
47	14	15	16	17	18	20
48	14	15	16	17	19	20
49	15	15	17	18	19	21
50	15	16	17	18	19	21

Table 30 SAMPLE SIZES FOR THE SIGN TEST
(Table: p. 91)

Let s be the critical value for the sign test, obtained in the same way as for Table 29 for some sample size n and significance level α. If the alternative is $p < 1/2$, then the power of the sign test against this alternative is given by

$$\text{power} = \sum_{j=0}^{s} \binom{n}{j} p^j (1 - p)^{n-j}$$

Table 30 contains, for given α, the smallest sample size n which attains the power given at the alternative p and the corresponding critical value s. The same n is appropriate for both p and $1 - p$, but s becomes $n - s$. Note that the type I error attained is less than or equal to α, and the true power is greater than or equal to the power indicated.

Table 30 SAMPLE SIZES FOR THE SIGN TEST
(Description: p. 90)

power = .80

p	α → 0.010 n	0.010 s	0.050 n	0.050 s	0.100 n	0.100 s	0.250 n	0.250 s
0.45	1167	539	786	365	620	289	398	187
0.40	291	123	199	85	158	63	103	45
0.35	131	50	90	35	69	27	45	18
0.30	73	25	49	17	37	13	25	9
0.25	44	13	30	9	23	7	16	5
0.20	32	8	20	5	18	5	11	3
0.15	21	4	15	3	13	3	9	2
0.10	15	2	10	2	8	1	6	1
0.05	12	1	9	1	5	0	4	0

power = .90

p	α → 0.010 n	0.010 s	0.050 n	0.050 s	0.100 n	0.100 s	0.250 n	0.250 s
0.45	1493	696	1055	495	866	408	591	281
0.40	373	161	263	115	213	94	147	66
0.35	164	65	114	46	93	38	67	28
0.30	92	33	65	24	53	20	36	14
0.25	57	18	42	14	33	11	23	8
0.20	39	11	28	8	23	7	16	5
0.15	26	6	17	4	16	4	11	3
0.10	18	3	15	3	11	2	9	2
0.05	15	2	9	1	8	1	6	1

power = .95

p	α → 0.010 n	0.010 s	0.050 n	0.050 s	0.100 n	0.100 s	0.250 n	0.250 s
0.45	1782	836	1302	615	1092	518	792	379
0.40	446	195	327	145	268	120	193	88
0.35	195	79	145	60	119	50	86	37
0.30	108	40	79	30	67	26	47	19
0.25	66	22	49	17	42	15	32	12
0.20	44	13	35	11	28	9	21	7
0.15	32	8	23	6	18	5	14	4
0.10	24	5	17	4	13	3	11	3
0.05	15	2	12	2	11	2	6	1

Table 31 CRITICAL VALUES FOR THE NUMBER OF RUNS TEST (Table: pp. 93-96)

Assume that m x's and n y's are arranged at random into a sequence (e.g., for m = 5, n = 6 one such sequence would be xxyxyyyxxyy). Let U be the number of distinct runs of x's and of y's in such a sequence (in our example U = 6). For given $m \leq n$, under the hypothesis that all arrangements are equally likely, the table contains the largest integer u such that $P\{U \leq u\} \leq \gamma$ for γ = 0.005, 0.01, 0.025, .05 and the smallest integer u such that $P\{U \leq u\} \geq \gamma$ for γ = 0.95, 0.975, 0.99, 0.995.

Table 31 CRITICAL VALUES FOR THE NUMBER OF RUNS TEST
(Description: p. 92)

		γ							
m	n	0.005	0.01	0.025	0.05	0.95	0.975	0.99	0.995
4	4	-	-	-	2	7	8	8	8
	5	-	-	2	2	8	8	8	9
	6	-	2	2	3	8	8	9	9
	7	-	2	2	3	8	9	9	9
	8	2	2	3	3	9	9	9	9
4	9	2	2	3	3	9	9	9	9
	10	2	2	3	3	9	9	9	9
5	5	-	2	2	3	8	9	9	10
	6	2	2	3	3	9	9	10	10
	7	2	2	3	3	9	10	10	11
	8	2	2	3	3	10	10	11	11
	9	2	3	3	4	10	11	11	11
5	10	3	3	3	4	10	11	11	11
6	6	2	2	3	3	10	10	11	11
	7	2	3	3	4	10	11	11	12
	8	3	3	3	4	11	11	12	12
	9	3	3	4	4	11	12	12	13
	10	3	3	4	5	11	12	13	13
6	11	3	4	4	5	12	12	13	13
	12	3	4	4	5	12	12	13	13
7	7	3	3	3	4	11	12	12	12
	8	3	3	4	4	12	12	13	13
	9	3	4	4	5	12	13	13	14
	10	3	4	5	5	12	13	14	14
	11	4	4	5	5	13	13	14	14
7	12	4	4	5	6	13	13	14	15
	13	4	5	5	6	13	14	15	15
	14	4	5	5	6	13	14	15	15
8	8	3	4	4	5	12	13	13	14
	9	3	4	5	5	13	13	14	14
	10	4	4	5	6	13	14	14	15
	11	4	5	5	6	14	14	15	15
	12	4	5	6	6	14	15	15	16
8	13	5	5	6	6	14	15	16	16
	14	5	5	6	7	15	15	16	16
	15	5	5	6	7	15	15	16	17
	16	5	6	6	7	15	16	16	17

Table 31 CRITICAL VALUES FOR THE NUMBER OF RUNS TEST (Description: p. 92)

		γ							
m	n	0.005	0.01	0.025	0.05	0.95	0.975	0.99	0.995
9	9	4	4	5	6	13	14	15	15
	10	4	5	5	6	14	15	15	16
	11	5	5	6	6	14	15	16	16
	12	5	5	6	7	15	15	16	17
	13	5	6	6	7	15	16	17	17
9	14	5	6	7	7	16	16	17	17
	15	6	6	7	8	16	17	17	18
	16	6	6	7	8	16	17	17	18
	17	6	7	7	8	16	17	18	18
	18	6	7	8	8	17	17	18	19
10	10	5	5	6	6	15	15	16	16
	11	5	5	6	7	15	16	17	17
	12	5	6	7	7	16	16	17	18
	13	5	6	7	8	16	17	18	18
	14	6	6	7	8	16	17	18	18
10	15	6	7	7	8	17	17	18	19
	16	6	7	8	8	17	18	19	19
	17	7	7	8	9	17	18	19	19
	18	7	7	8	9	18	18	19	20
	19	7	8	8	9	18	19	19	20
10	20	7	8	9	9	18	19	19	20
11	11	5	6	7	7	16	16	17	18
	12	6	6	7	8	16	17	18	18
	13	6	6	7	8	17	18	18	19
	14	6	7	8	8	17	18	19	19
	15	7	7	8	9	18	18	19	20
11	16	7	7	8	9	18	19	20	20
	17	7	8	9	9	18	19	20	21
	18	7	8	9	10	19	19	20	21
	19	8	8	9	10	19	20	21	21
	20	8	8	9	10	19	20	21	21
12	12	6	7	7	8	17	18	18	19
	13	6	7	8	9	17	18	19	20
	14	7	7	8	9	18	19	20	20
	15	7	8	8	9	18	19	20	21
	16	7	8	9	10	19	20	21	21
12	17	8	8	9	10	19	20	21	21
	18	8	8	9	10	20	20	21	22
	19	8	9	10	10	20	21	22	22
	20	8	9	10	11	20	21	22	22

Table 31 CRITICAL VALUES FOR THE NUMBER OF RUNS TEST
(Description: p. 92)

m	n	γ 0.005	0.01	0.025	0.05	0.95	0.975	0.99	0.995
13	13	7	7	8	9	18	19	20	20
	14	7	8	9	9	19	19	20	21
	15	7	8	9	10	19	20	21	21
	16	8	8	9	10	20	20	21	22
	17	8	9	10	10	20	21	22	22
13	18	8	9	10	11	20	21	22	23
	19	9	9	10	11	21	22	23	23
	20	9	10	10	11	21	22	23	23
14	14	7	8	9	10	19	20	21	22
	15	8	8	9	10	20	21	22	22
	16	8	9	10	11	20	21	22	23
	17	8	9	10	11	21	22	23	23
	18	9	9	10	11	21	22	23	24
14	19	9	10	11	12	22	22	23	24
	20	9	10	11	12	22	23	24	24
15	15	8	9	10	11	20	21	22	23
	16	9	9	10	11	21	22	23	23
	17	9	10	11	11	21	22	23	24
	18	9	10	11	12	22	23	24	24
	19	10	10	11	12	22	23	24	25
15	20	10	11	12	12	23	24	25	25
16	16	9	10	11	11	22	22	23	24
	17	9	10	11	12	22	23	24	25
	18	10	10	11	12	23	24	25	25
	19	10	11	12	13	23	24	25	26
	20	10	11	12	13	24	24	25	26
17	17	10	10	11	12	23	24	25	25
	18	10	11	12	13	23	24	25	26
	19	10	11	12	13	24	25	26	26
	20	11	11	13	13	24	25	26	27
18	18	11	11	12	13	24	25	26	26
	19	11	12	13	14	24	25	26	27
	20	11	12	13	14	25	26	27	28
19	19	11	12	13	14	25	26	27	28
	20	12	12	13	14	26	26	28	28
20	20	12	13	14	15	26	27	28	29

Table 32 DISTRIBUTION OF THE MAGNITUDE OF QUADRANT SUMS
(Table: p. 97)

The quadrant test is a nonparametric test of association that places weight on the extreme values. Given a bivariate sample (X_i, Y_i), $i = 1, 2, \ldots, n$, with no ties, let X_m and Y_m be the X and Y sample medians. Let $\bar{Y}_{max}$ be the Y-value of the sample point that has the maximum X-value. Similarly, define $\bar{Y}_{min}$, $\bar{X}_{min}$, $\bar{X}_{max}$. Create a sequence of +'s and -'s by ordering the sample points by the X_i values and recording a + if and only if the Y_i values are greater than or equal to Y_m. Take the first and last runs and let ℓ_1 be the length of the first run (containing a + or - from the point with minimum X-value) and ℓ_2 be the length of the last run (maximum X). Similarly define, by interchanging the roles of X and Y, ℓ_3 (minimum Y) and ℓ_4 (maximum Y). Let $\text{sgn}(a) = 1$ if $a \geq 0$, -1 if $a < 0$. The quadrant sum is $S = \text{sgn}(Y_m - \bar{Y}_{min})\ell_1 + \text{sgn}(\bar{Y}_{max} - Y_m)\ell_2 + \text{sgn}(X_m - \bar{X}_{min})\ell_3 + \text{sgn}(\bar{X}_{max} - X_m)\ell_4$. The table presents $P(|S| \geq s)$ under the hypothesis that the X and Y values are independent.

Table 32 DISTRIBUTION OF THE MAGNITUDE OF QUADRANT SUMS
(Description: p. 96)

s	sample size, n						
	6	8	10	12	14	16	∞
0	1.0000	1.0000	1.0000	1.0000	1.0000	1.0000	1.0000
1	0.9333	0.9036	0.9106	0.9108	0.9115	0.9118	0.9120
2	0.7556	0.7544	0.7567	0.7576	0.7580	0.7583	0.7546
3	0.6000	0.6000	0.6008	0.6030	0.6039	0.6044	0.5996
4	0.4667	0.4619	0.4662	0.4679	0.4690	0.4695	0.4630
5	0.3111	0.3508	0.3519	0.3537	0.3647	0.3551	0.3469
6	0.2222	0.2619	0.2589	0.2604	0.2611	0.2613	0.2520
7	0.1556	0.1821	0.1867	0.1873	0.1876	0.1876	0.1777
8	0.1111	0.1258	0.1333	0.1322	0.1322	0.1320	0.1218
9	0.1000	0.0839	0.0928	0.0922	0.0918	0.0914	0.0815
10	0.1000	0.0554	0.0642	0.0640	0.0632	0.0626	0.0533
11	0.1000	0.0375	0.0436	0.0442	0.0432	0.0425	0.0342
12	0.1000	0.0304	0.0290	0.0305	0.0296	0.0288	0.0216
13	0.0000	0.0286	0.0190	0.0209	0.0202	0.0195	0.0134
14		0.0286	0.0127	0.0143	0.0139	0.0132	0.0082
15		0.0286	0.0095	0.0096	0.0096	0.0090	0.0050
16		0.0286	0.0083	0.0064	0.0066	0.0062	0.0030
17		0.0000	0.0079	0.0043	0.0045	0.0042	0.0018
18			0.0079	0.0031	0.0031	0.0029	0.0010
19			0.0079	0.0025	0.0021	0.0020	0.0006
20			0.0079	0.0022	0.0014	0.0014	0.0003
21			0.0000	0.0022	0.0010	0.0010	0.0002
22				0.0022	0.0008	0.0007	0.0001
23				0.0022	0.0006	0.0005	0.0001
24				0.0022	0.0006	0.0003	0.0000
25				0.0000	0.0006	0.0002	
26					0.0006	0.0002	
27					0.0006	0.0002	
28					0.0006	0.0002	
29					0.0000	0.0002	
30						0.0002	
31						0.0002	
32						0.0002	

RANK CORRELATION (Tables 33 and 34)

Consider a bivariate sample of n pairs (X_i, Y_i). Suppose there is zero probability of ties. Rank the X's among themselves and the Y's among themselves, giving n pairs of ranks $(R1_i, R2_i)$. Tables 33 and 34 give critical values for testing the null hypothesis that the X and Y variables are independent.

Table 33 CRITICAL VALUES FOR KENDALL'S RANK CORRELATION COEFFICIENT

(Table: p. 99)

For each i, i = 1, 2, ..., n, count the number of pairs $(R1_j, R2_j)$ with $R1_j > R1_i$ and $R2_j > R2_i$. Let K be the sum of these n counts. For each cumulative probability γ, the table contains the smallest integer k such that $P(K \geq k) \leq 1 - \gamma$. Since the statistic is symmetric upon interchange of the roles of X and Y the table may be used to construct two-sided tests. Kendall's rank correlation coefficient is $\zeta = 4K/\{n(n - 1)\} - 1$.

Table 34 CRITICAL VALUES FOR SPEARMAN'S RANK CORRELATION COEFFICIENT

(Table: p. 100)

Let $S = \sum_{i=1}^{n}(R1_i - R2_i)^2$. Tabulated for given n are pairs s and $P(S \leq s)$ under independence. The second entry of the pair, the probability, is given in parentheses. The mean of S is $(n^3 - n)/6$ and the variance is $n^2(n + 1)^2(n - 1)/36$. Spearman's rank correlation coefficient is $1 - 6S/(n^3 - n)$.

Table 33 CRITICAL VALUES FOR KENDALL'S RANK CORRELATION COEFFICIENT
(Description: p. 98)

					γ					
n	.70	.75	.80	.85	.90	.95	.99	.975	.995	.999
3	3	3	3	-	-	-	-	-	-	-
4	5	5	5	6	6	6	-	-	-	-
5	7	7	8	8	9	9	10	10	-	-
6	10	10	11	11	12	13	14	14	15	-
7	13	14	14	15	16	17	18	19	20	21
8	17	18	18	19	20	22	23	24	25	26
9	22	22	23	24	25	27	28	30	31	33
10	27	27	28	29	31	33	34	36	37	40
11	32	33	34	35	37	39	41	43	44	47
12	38	39	40	42	43	46	48	51	52	55
13	44	46	47	49	51	53	56	59	61	64
14	51	53	54	56	58	62	64	67	69	73
15	59	60	62	64	67	70	73	77	79	83
16	67	69	70	73	75	79	83	86	89	94
17	75	77	79	82	85	89	93	97	100	105
18	85	87	89	91	95	99	103	108	111	117
19	94	96	99	101	105	110	114	119	123	129
20	104	107	109	112	116	121	126	131	135	142
21	115	117	120	123	127	133	138	144	148	156
22	126	129	132	135	139	146	151	157	161	170
23	138	140	144	147	152	159	164	171	176	184
24	150	153	156	160	165	172	178	185	190	200
25	162	166	169	173	179	186	193	200	205	216

Table 34 CRITICAL VALUES FOR SPEARMAN'S RANK CORRELATION COEFFICIENT
(Description: p. 98)

n = 3

0 (0.167) 2 (0.500)

n = 4

0 (0.042) 2 (0.167) 4 (0.208) 6 (0.375) 8 (0.458)

n = 5

0 (0.008) 2 (0.042) 4 (0.067) 6 (0.117) 8 (0.175)
10 (0.225) 12 (0.258) 14 (0.342) 16 (0.392) 18 (0.475)

n = 6

0 (0.001) 2 (0.008) 4 (0.017) 6 (0.029) 8 (0.051)
10 (0.068) 12 (0.087) 14 (0.121) 18 (0.178) 20 (0.210)

n = 7

2 (0.001) 4 (0.003) 6 (0.006) 8 (0.012) 12 (0.024)
14 (0.033) 16 (0.044) 18 (0.055) 22 (0.083) 24 (0.100)
34 (0.198) 36 (0.222)

n = 8

4 (0.001) 6 (0.001) 10 (0.004) 12 (0.005) 14 (0.008)
16 (0.011) 22 (0.023) 24 (0.029) 30 (0.048) 32 (0.057)
40 (0.098) 42 (0.108) 54 (0.195) 56 (0.214)

n = 9

10 (0.001) 12 (0.001) 20 (0.004) 22 (0.005) 26 (0.009)
28 (0.011) 36 (0.022) 38 (0.025) 48 (0.048) 50 (0.054)
62 (0.097) 64 (0.106) 80 (0.193) 82 (0.205)

n = 10

20 (0.001) 22 (0.001) 34 (0.004) 36 (0.005) 42 (0.009)
44 (0.010) 58 (0.024) 60 (0.027) 72 (0.048) 74 (0.052)
90 (0.096) 92 (0.102) 114 (0.193) 116 (0.203)

NONPARAMETRIC ONE WAY ANALYSIS OF VARIANCE
(Tables 35, 36, 37 and 38)

Consider observations from k groups, $j = 1, \ldots, k$, the jth group having n_j observations. The model is

$$X_{ij} = \mu_j + e_{ij};\ i = 1, \ldots, n_j,\ j = 1, 2, \ldots, k$$

where the μ_j are constants and the e_{ij} are independent and identically distributed errors from a continuous distribution. Let $N = \sum_{j=1}^{k} n_j$ be the total number of observations, and let r_{ij} be the rank of X_{ij} when jointly ranking all N variables. The usual null hypothesis is H_0: $\mu_1 = \mu_2 = \cdots = \mu_k$. Let $R_j = \sum_{i=1}^{n_j} r_{ij}$.

Table 35 CRITICAL VALUES FOR THE KRUSKAL-WALLIS STATISTIC (Table: pp. 104-105)

For testing H_0 against the alternative that H_0 does not hold, the Kruskal-Wallis statistic is

$$T = \frac{12\{\sum_{j=1}^{k}(R_j^2/n_j)\}}{N(N+1)} - 3(N+1)$$

For k groups of sizes $n_1, n_2, \ldots, n_k$, critical values for T, say t, are given, followed in parentheses by $P(T \geq t | H_0 \text{ holds})$.

Table 36 CRITICAL VALUES FOR JONCKHEERE'S STATISTIC (Table: pp. 106-108)

The notation for this table is given in the general description for this group of tables. Jonckheere's statistic is designed for power against the alternative, H_A: $\mu_1 \leq \mu_2 \leq \cdots \leq \mu_k$ with at least one strict inequality. Let $\emptyset(a, b) = 1$ if $a < b$, $= 0$ otherwise. Jonckheere's statistic is

$$J = \sum_{r=1}^{k-1} \sum_{s=r+1}^{k} \sum_{i=1}^{n_r} \sum_{m=1}^{n_s} \emptyset(X_{ir}, X_{ms})$$

That is, for two groups labeled by $r < s$, J counts the number of times a value in group r is less than a value in group s. On the first two pages of the table, critical values j are tabulated, for given combinations of $n_1, \ldots, n_k$, $k = 3$ or 4, and α. Tabled is the smallest value of j such that $P(J \geq j | H_0 \text{ holds}) \leq \alpha$. The last page of the table considers k groups with $n_1 = n_2 = \cdots = n_k = n$. For given combinations of k, n and α, critical values j are tabulated such that $P(J \geq j | H_0 \text{ holds}) \leq \alpha$.

Table 37 MULTIPLE COMPARISONS BASED ON KRUSKAL-WALLIS RANK SUMS (Table: p. 109)

The notation for this table is given in the general description for this group of tables. Consider k groups of size n, that is, $n_1 = n_2 = n_3 = \cdots = n_k = n$. Compute $D = \max_{i,j} |R_i - R_j|$. Tabulated for given k and n are critical values of D, say d, followed in parentheses by $P(D \geq d | H_0 \text{ holds})$.

Thus, one may compare all pairs R_i and R_j and decide that $\mu_i \neq \mu_j$ whenever $|R_i - R_j| \geq d$ with an experimentwise error rate, as given in parentheses, of falsely detecting a difference (under H_0).

Table 38 CRITICAL VALUES OF THE MAXIMUM RANK SUM
(Table: p. 110)

The notation for this table is given in the general description for this group of tables. Consider k groups of size n and let $M = \max R_i$. Tabulated for indicated values of k and n are values m, followed in parentheses by $P(M \geq m | H_0$ holds). If $M \geq m$ one concludes that the group i for which $R_i = M$ has the largest mean of the k groups.

Table 35 CRITICAL VALUES FOR THE KRUSKAL-WALLIS STATISTIC (Description: pp 101-103)

$n_1 = 3, n_2 = 3, n_3 = 3$

7.200 (0.004) 6.489 (0.011) 5.956 (0.025) 5.600 (0.050)
5.422 (0.071) 4.622 (0.100) 4.356 (0.132) 3.467 (0.196)
3.289 (0.232)

$n_1 = 4, n_2 = 3, n_3 = 3$

8.018 (0.001) 7.318 (0.004) 7.000 (0.006) 6.745 (0.010)
6.709 (0.013) 6.155 (0.025) 6.018 (0.027) 5.791 (0.046)
5.727 (0.050) 4.709 (0.092) 4.700 (0.101) 3.391 (0.196)
3.364 (0.203)

$n_1 = 4, n_2 = 4, n_3 = 3$

8.909 (0.001) 8.326 (0.001) 7.598 (0.004) 7.477 (0.006)
7.144 (0.010) 7.136 (0.011) 6.394 (0.025) 6.386 (0.026)
5.598 (0.049) 5.576 (0.051) 4.545 (0.099) 4.477 (0.102)
3.417 (0.195) 3.394 (0.201)

$n_1 = 4, n_2 = 4, n_3 = 4$

9.269 (0.001) 8.769 (0.001) 8.000 (0.005) 7.731 (0.007)
7.654 (0.008) 7.538 (0.011) 6.615 (0.024) 6.577 (0.026)
5.692 (0.049) 5.654 (0.055) 4.654 (0.097) 4.500 (0.104)
3.500 (0.197) 3.231 (0.212)

$n_1 = 5, n_2 = 4, n_3 = 4$

9.168 (0.001) 9.129 (0.001) 8.189 (0.005) 8.156 (0.005)
7.760 (0.009) 7.744 (0.011) 6.673 (0.024) 6.597 (0.026)
5.657 (0.049) 5.618 (0.050) 4.668 (0.098) 4.619 (0.100)
3.382 (0.197) 3.330 (0.200)

$n_1 = 5, n_2 = 5, n_3 = 4$

9.606 (0.001) 9.506 (0.001) 8.523 (0.005) 8.463 (0.005)
7.823 (0.010) 7.791 (0.010) 6.760 (0.025) 6.671 (0.025)
5.666 (0.049) 5.643 (0.050) 4.523 (0.099) 4.520 (0.101)
3.311 (0.200) 3.286 (0.203)

$n_1 = 5, n_2 = 5, n_3 = 5$

9.920 (0.001) 9.780 (0.001) 8.780 (0.005) 8.720 (0.005)
8.000 (0.009) 7.980 (0.011) 6.740 (0.025) 6.720 (0.026)
5.780 (0.049) 5.660 (0.051) 4.560 (0.100) 4.500 (0.102)
3.420 (0.190) 3.380 (0.201)

Table 35 CRITICAL VALUES FOR THE KRUSKAL-WALLIS STATISTIC (Description: pp. 101-103)

$n_1 = 6,\ n_2 = 5,\ n_3 = 5$

10.287 (0.001)	10.271 (0.001)	8.859 (0.005)	8.835 (0.005)
8.028 (0.010)	8.012 (0.010)	6.788 (0.025)	6.781 (0.025)
5.729 (0.050)	5.699 (0.051)	4.547 (0.098)	4.529 (0.102)
3.282 (0.200)	3.276 (0.202)		

$n_1 = 6,\ n_2 = 6,\ n_3 = 5$

10.524 (0.001)	10.515 (0.001)	8.987 (0.005)	8.982 (0.005)
8.124 (0.010)	8.119 (0.010)	6.848 (0.025)	6.838 (0.025)
5.765 (0.050)	5.752 (0.050)	4.542 (0.100)	4.541 (0.101)
3.293 (0.200)	3.282 (0.202)		

$n_1 = 6,\ n_2 = 6,\ n_3 = 6$

10.889 (0.001)	10.842 (0.001)	9.170 (0.005)	9.088 (0.005)
8.222 (0.010)	8.187 (0.010)	6.889 (0.025)	6.877 (0.026)
5.801 (0.049)	5.719 (0.050)	4.643 (0.099)	4.538 (0.101)
3.310 (0.196)	3.263 (0.200)		

$n_1 = 2,\ n_2 = 2,\ n_3 = 2,\ n_4 = 2$

6.667 (0.010)	6.167 (0.038)	6.000 (0.067)	5.667 (0.076)
5.500 (0.114)	4.833 (0.171)	4.667 (0.210)	

$n_1 = 3,\ n_2 = 3,\ n_3 = 3,\ n_4 = 3$

9.513 (0.001)	9.462 (0.001)	8.897 (0.004)	8.744 (0.006)
8.538 (0.008)	8.436 (0.011)	7.667 (0.023)	7.615 (0.026)
7.000 (0.044)	6.897 (0.050)	6.026 (0.098)	5.974 (0.103)
4.795 (0.196)	4.744 (0.209)		

$n_1 = 4,\ n_2 = 4,\ n_3 = 4,\ n_4 = 4$

11.360 (0.001)	11.338 (0.001)	9.971 (0.005)	9.949 (0.005)
9.287 (0.010)	9.265 (0.010)	8.228 (0.025)	8.206 (0.025)
7.235 (0.049)	7.213 (0.051)	6.088 (0.099)	6.066 (0.100)
4.721 (0.198)	4.699 (0.202)		

$n_1 = 2,\ n_2 = 2,\ n_3 = 2,\ n_4 = 2,\ n_5 = 2$

8.727 (0.001)	8.400 (0.005)	8.291 (0.010)	8.073 (0.013)
7.964 (0.022)	7.855 (0.025)	7.418 (0.049)	7.309 (0.063)
6.982 (0.091)	6.873 (0.102)	5.891 (0.199)	5.782 (0.220)

$n_1 = 3,\ n_2 = 3,\ n_3 = 3,\ n_4 = 3,\ n_5 = 3$

11.667 (0.001)	11.633 (0.001)	10.733 (0.005)	10.700 (0.005)
10.200 (0.010)	10.167 (0.010)	9.233 (0.025)	9.200 (0.025)
8.333 (0.050)	8.300 (0.051)	7.333 (0.099)	7.300 (0.101)
6.067 (0.197)	6.033 (0.200)		

Table 36 CRITICAL VALUES FOR JONCKHEERE'S STATISTIC
(Description: pp. 101-103)

n_1	n_2	n_3	.20	.10	.05	.025	.01	.005
			significance level, α					
3	3	3	18	20	22	23	25	25
3	3	4	22	24	26	28	29	30
3	3	5	26	28	30	32	34	35
3	4	4	26	29	31	33	35	36
3	4	5	30	33	36	38	40	41
3	5	5	35	38	41	43	46	47
4	4	4	31	34	36	38	40	42
4	4	5	36	39	42	44	46	48
4	4	6	40	44	47	49	52	54
4	5	5	41	45	48	50	53	55
4	5	6	46	50	54	56	59	62
4	6	6	52	56	60	63	67	69
5	5	5	47	51	54	57	60	62
5	5	6	52	57	61	64	67	70
5	5	7	58	63	67	71	74	77
5	6	6	59	64	68	71	75	77
5	6	7	65	70	75	79	83	85
5	7	7	72	78	82	86	91	94
6	6	6	66	71	75	79	83	86
6	6	7	72	78	83	87	92	95
6	6	8	79	86	91	95	100	103
6	7	7	80	86	91	96	101	104
6	7	8	87	94	100	104	110	113
6	8	8	95	102	108	113	119	123
7	7	7	88	95	100	105	110	114
7	7	8	96	103	109	114	120	123
7	7	9	104	111	118	123	129	133
7	8	8	104	112	118	124	130	134
7	8	9	113	121	128	133	140	144
7	9	9	122	130	137	144	151	155
8	8	8	113	121	128	134	140	145
8	8	9	122	131	138	144	151	156
8	8	10	131	140	148	154	162	167
8	9	9	132	141	149	155	162	167
8	9	10	141	151	159	166	174	179
8	10	10	151	162	170	177	186	191
9	9	9	142	152	160	166	174	180
9	9	10	152	162	171	178	186	192
9	10	10	162	173	182	190	199	204
10	10	10	173	185	194	202	212	218

Table 36 CRITICAL VALUES FOR JONCKHEERE'S STATISTIC (Description: pp. 101-103)

n_1	n_2	n_3	n_4	.20	.10	.05	.025	.01	.005
				significance level, α					
3	3	3	3	34	37	40	42	44	45
3	3	4	4	45	49	52	55	58	59
3	4	4	4	51	56	59	62	65	67
4	4	4	4	58	63	67	70	73	76
4	4	5	5	72	78	82	86	91	94
4	5	5	5	80	86	91	95	100	103
5	5	5	5	89	95	100	105	110	114
5	5	6	6	106	114	120	125	131	135
5	6	6	6	116	124	130	136	142	146
6	6	6	6	126	134	141	147	154	158
6	6	7	7	146	156	164	170	178	183
6	7	7	7	157	167	176	183	191	196
7	7	7	7	169	179	188	196	204	210
7	7	8	8	192	204	214	222	232	238
7	8	8	8	205	218	228	236	246	253
8	8	8	8	218	231	242	251	262	269
8	8	9	9	245	259	271	281	293	300
8	9	9	9	259	274	286	297	309	317
9	9	9	9	274	290	302	313	326	334
9	9	10	10	304	321	335	347	360	369
9	10	10	10	320	337	352	364	378	388
10	10	10	10	336	354	369	382	397	407

Table 36 CRITICAL VALUES FOR JONCKHEERE'S STATISTIC
(Description: pp. 101-103)

	n	significance level, α .20	.10	.05	.025	.01	.005
k = 5	3	54	59	62	65	69	71
	4	94	100	106	110	116	119
	5	144	153	160	167	174	179
	6	204	216	226	235	244	251
	7	275	290	303	313	325	334
	8	357	375	390	403	418	428
	9	448	470	488	504	522	534
	10	550	576	597	615	636	650
k = 6	3	80	85	90	94	98	101
	4	138	147	154	160	167	171
	5	212	224	234	242	252	259
	6	302	317	330	342	354	363
	7	407	427	443	457	474	484
	8	528	552	572	589	609	623
	9	664	693	717	738	761	777
	10	816	850	878	902	930	949
k = 7	3	109	117	122	127	133	137
	4	190	201	210	218	227	233
	5	293	308	321	332	344	352
	6	418	438	454	468	484	495
	7	564	589	610	628	648	662
	8	732	763	788	810	835	852
	9	922	959	989	1015	1045	1065
	10	1133	1176	1211	1242	1277	1301
k = 8	3	144	153	160	166	173	178
	4	251	264	275	285	296	303
	5	387	406	421	434	449	460
	6	552	577	597	614	634	648
	7	746	777	802	824	849	867
	8	969	1007	1038	1064	1095	1116
	9	1221	1266	1303	1335	1371	1396
	10	1502	1554	1597	1635	1678	1707

Table 37 MULTIPLE COMPARISONS BASED ON KRUSKAL-WALLIS RANK SUMS (Description: pp. 101-103)

k = 3, n = 3

18 (0.004)   17 (0.011)   16 (0.029)   15 (0.064)   14 (0.111)
13 (0.161)   12 (0.218)

k = 3, n = 4

31 (0.001)   30 (0.001)   29 (0.003)   28 (0.007)   27 (0.011)
26 (0.020)   25 (0.031)   24 (0.045)   23 (0.062)   22 (0.083)
21 (0.106)   19 (0.169)   18 (0.207)

k = 3, n = 5

45 (0.001)   44 (0.001)   41 (0.004)   40 (0.006)   39 (0.009)
38 (0.013)   36 (0.023)   35 (0.030)   33 (0.048)   32 (0.059)
30 (0.088)   29 (0.106)   25 (0.199)   24 (0.229)

k = 3, n = 6

60 (0.001)   59 (0.001)   55 (0.004)   54 (0.005)   52 (0.009)
51 (0.011)   47 (0.025)   46 (0.030)   43 (0.049)   42 (0.058)
39 (0.090)   38 (0.103)   33 (0.188)   32 (0.210)

k = 4, n = 2

12 (0.029)   11 (0.086)   10 (0.210)

k = 4, n = 3

27 (0.001)   26 (0.002)   25 (0.005)   24 (0.012)   23 (0.023)
22 (0.042)   21 (0.072)   20 (0.108)   19 (0.154)   18 (0.204)

k = 4, n = 4

43 (0.001)   42 (0.002)   40 (0.005)   39 (0.008)   38 (0.012)
36 (0.025)   35 (0.034)   34 (0.046)   33 (0.060)   31 (0.097)
30 (0.119)   28 (0.173)   27 (0.204)

k = 5, n = 2

16 (0.016)   15 (0.048)   14 (0.121)   13 (0.222)

k = 5, n = 3

35 (0.001)   34 (0.002)   33 (0.004)   32 (0.007)   31 (0.013)
30 (0.023)   29 (0.037)   28 (0.057)   27 (0.084)   26 (0.117)
25 (0.157)   24 (0.201)

Table 38 CRITICAL VALUES OF THE MAXIMUM RANK SUM
(Description: pp. 101-103)

k = 3, n = 3

24 (0.036) 23 (0.071) 22 (0.143) 21 (0.250)

k = 3, n = 4

42 (0.006) 41 (0.012) 40 (0.024) 39 (0.042) 38 (0.073)
37 (0.109) 36 (0.164) 35 (0.230)

k = 3, n = 5

65 (0.001) 63 (0.004) 62 (0.007) 61 (0.012) 60 (0.019)
59 (0.029) 58 (0.042) 57 (0.060) 56 (0.083) 55 (0.113)
53 (0.194) 52 (0.247)

k = 3, n = 6

91 (0.001) 90 (0.001) 87 (0.005) 86 (0.007) 85 (0.010)
83 (0.020) 82 (0.027) 80 (0.048) 79 (0.062) 78 (0.079)
77 (0.100) 74 (0.187) 73 (0.226)

k = 4, n = 2

15 (0.143) 14 (0.286)

k = 4, n = 3

33 (0.018) 32 (0.036) 31 (0.073) 30 (0.127) 29 (0.200)

k = 4, n = 4

58 (0.002) 57 (0.004) 56 (0.009) 55 (0.015) 54 (0.026)
53 (0.040) 52 (0.059) 51 (0.084) 50 (0.116) 49 (0.156)
48 (0.205)

k = 5, n = 2

19 (0.111) 18 (0.222)

k = 5, n = 3

42 (0.011) 41 (0.022) 40 (0.044) 39 (0.077) 38 (0.121)
37 (0.175) 36 (0.249)

NONPARAMETRIC TWO-WAY ANALYSIS OF VARIANCE
(Tables 39, 40 and 41)

The data consist of observations in n "blocks," each block having one observation from each of k "treatments." The situation is modeled:

$$X_{ij} = \mu + \beta_i + \tau_j + \varepsilon_{ij}; \; i = 1, \ldots, n, \quad j = 1, \ldots, k$$

where $\sum_{i=1}^{n} \beta_i = \sum_{j=1}^{n} \tau_j = 0$ and the nk ε_{ij}'s are independent, continuous identically distributed random variables. The usual null hypothesis of interest is H_0: $\tau_1 = \tau_2 = \cdots = \tau_k$ (= 0), that is, there is no treatment effect. Within each block rank the X_{ij}'s and let r_{ij} be the rank of X_{ij} among the ranking of $X_{i1}, \ldots, X_{ik}$. Let $R_j = \sum_{i=1}^{n} r_{ij}$.

Table 39 CRITICAL VALUES FOR FRIEDMAN'S CHI SQUARE STATISTIC (Table: pp. 113-116)

Friedman's chi square statistic is given by

$$T = \frac{12 \sum_{j=1}^{k} R_j^2}{nk(k+1)} - 3n(k+1)$$

Tabulated for given k and n are critical values t of T, followed in parentheses by $P(T \geq t | H_0 \text{ holds})$.

As n approaches infinity T is asymptotically χ^2 with k - 1 degrees of freedom if H_0 holds. The design of T is to give good power against a general alternative.

Table 40 CRITICAL VALUES FOR PAGE'S L STATISTIC (Table: p. 117)

Page's L statistic, $L = \sum_{j=1}^{k} jR_j$, is designed to give good power for the ordered alternative H_A: $\tau_1 \leq \tau_2 \leq \cdots \leq \tau_k$ (notation given in the general description for this group of tables). For given triples (significance level) α, k and n, tabulated are the smallest ℓ for which $P(L \geq \ell | H_0 \text{ holds}) \leq \alpha$.

Table 41 MULTIPLE COMPARISONS BASED ON FRIEDMAN RANK SUMS (Table: pp. 118-119)

The notation for this table is given in the general description for this group of tables. Let $R = \max_{i,i'} |R_i - R_{i'}|$. That is, D is the maximum absolute deviation of the rank sums among the k groups. Tabulated for given k, n and r (possible values of R) are $P(R \geq r | H_0 \text{ holds})$. Thus, one may test all pairs of means for differences by rejecting $\tau_i = \tau_{i'}$ when $|R_i - R_{i'}| \geq r$ (as tabled). The probability (experimentwise error rate) of falsely detecting one or more pairs as different under H_0 is the tabled probability.

Table 39 CRITICAL VALUES FOR FRIEDMAN'S CHI SQUARE STATISTIC (Description: pp. 111-112)

k = 3, n = 2

4.000 (0.167) 3.000 (0.500)

k = 3, n = 3

6.000 (0.028) 4.667 (0.194) 2.667 (0.361)

k = 3, n = 4

8.000 (0.005) 6.500 (0.042) 6.000 (0.069) 4.500 (0.125)
3.500 (0.273)

k = 3, n = 5

10.000 (0.001) 8.400 (0.008) 7.600 (0.024) 6.400 (0.039)
5.200 (0.093) 4.800 (0.124) 3.600 (0.182) 2.800 (0.367)

k = 3, n = 6

12.000 (0.000) 10.333 (0.002) 9.333 (0.006) 9.000 (0.008)
8.333 (0.012) 7.000 (0.029) 6.333 (0.052) 5.333 (0.072)
4.333 (0.142) 4.000 (0.184) 3.000 (0.252)

k = 3, n = 7

12.286 (0.000) 11.143 (0.001) 10.286 (0.004) 8.857 (0.008)
8.000 (0.016) 7.714 (0.021) 7.143 (0.027) 6.000 (0.051)
5.429 (0.085) 4.571 (0.112) 3.714 (0.192) 3.429 (0.237)

k = 3, n = 8

12.250 (0.001) 12.000 (0.001) 9.750 (0.005) 9.000 (0.010)
7.750 (0.018) 7.000 (0.030) 6.250 (0.047) 5.250 (0.079)
4.750 (0.120) 4.000 (0.149) 3.250 (0.236)

Table 39 CRITICAL VALUES FOR FRIEDMAN'S CHI SQUARE STATISTIC (Description: pp. 111-113)

k = 4, n = 2

6.000 (0.042)	5.400 (0.167)	4.800 (0.208)	

k = 4, n = 3

9.000 (0.002)	8.200 (0.017)	7.400 (0.033)	7.000 (0.054)
6.600 (0.075)	5.800 (0.148)	5.400 (0.175)	5.000 (0.207)

k = 4, n = 4

11.100 (0.001)	10.800 (0.002)	10.200 (0.003)	9.900 (0.006)
9.600 (0.007)	9.300 (0.012)	8.400 (0.019)	8.100 (0.033)
7.800 (0.036)	7.500 (0.052)	6.300 (0.094)	6.000 (0.105)
5.100 (0.190)	4.800 (0.200)		

k = 4, n = 5

12.600 (0.001)	12.120 (0.001)	10.920 (0.003)	10.680 (0.005)
9.960 (0.009)	9.720 (0.012)	8.760 (0.023)	8.280 (0.031)
7.800 (0.044)	7.320 (0.055)	6.360 (0.093)	6.120 (0.107)
5.160 (0.162)	4.920 (0.210)		

k = 4, n = 6

12.800 (0.001)	12.600 (0.001)	11.400 (0.004)	11.000 (0.006)
10.200 (0.010)	10.000 (0.010)	8.800 (0.023)	8.600 (0.029)
7.600 (0.043)	7.400 (0.056)	6.400 (0.089)	6.200 (0.108)
4.800 (0.197)	4.600 (0.218)		

k = 4, n = 7

13.457 (0.001)	13.286 (0.001)	11.400 (0.004)	11.229 (0.005)
10.543 (0.009)	10.371 (0.010)	9.000 (0.023)	8.657 (0.030)
7.800 (0.041)	7.629 (0.052)	6.429 (0.093)	6.257 (0.100)
4.886 (0.195)	4.543 (0.220)		

k = 4, n = 8

13.800 (0.001)	13.650 (0.001)	11.550 (0.005)	11.400 (0.005)
10.500 (0.009)	10.350 (0.011)	9.000 (0.023)	8.850 (0.025)
7.650 (0.049)	7.500 (0.051)	6.300 (0.100)	6.150 (0.106)
4.800 (0.192)	4.650 (0.219)		

Table 39 CRITICAL VALUES FOR FRIEDMAN'S CHI SQUARE STATISTIC (Description: pp. 111-112)

k = 5, n = 2

8.000 (0.008)	7.600 (0.042)	7.200 (0.067)	6.800 (0.117)
6.400 (0.175)	6.000 (0.225)		

k = 5, n = 3

11.467 (0.001)	10.667 (0.004)	10.400 (0.005)	10.133 (0.008)
9.867 (0.015)	9.600 (0.017)	9.333 (0.026)	8.533 (0.045)
8.267 (0.056)	7.467 (0.096)	7.200 (0.117)	6.400 (0.172)
6.133 (0.213)			

k = 5, n = 4

13.200 (0.001)	13.000 (0.001)	12.000 (0.004)	11.800 (0.005)
11.200 (0.008)	11.000 (0.010)	9.800 (0.025)	9.600 (0.028)
8.800 (0.049)	8.600 (0.060)	7.600 (0.095)	7.400 (0.113)
6.200 (0.197)	6.000 (0.205)		

k = 5, n = 5

14.400 (0.001)	14.240 (0.001)	12.480 (0.005)	12.320 (0.006)
11.680 (0.009)	11.520 (0.010)	10.240 (0.024)	10.080 (0.026)
8.960 (0.049)	8.800 (0.056)	7.680 (0.094)	7.520 (0.107)
6.080 (0.195)	5.920 (0.218)		

k = 5, n = 6

15.200 (0.001)	15.067 (0.001)	13.067 (0.004)	12.933 (0.005)
11.867 (0.010)	11.733 (0.011)	10.400 (0.024)	10.267 (0.027)
9.067 (0.049)	8.933 (0.055)	7.733 (0.095)	7.600 (0.102)
6.133 (0.192)	6.000 (0.204)		

k = 5, n = 7

15.657 (0.001)	15.543 (0.001)	13.257 (0.005)	13.143 (0.005)
12.114 (0.010)	12.000 (0.011)	10.514 (0.025)	10.400 (0.026)
9.143 (0.049)	9.029 (0.053)	7.771 (0.094)	7.657 (0.103)
6.171 (0.187)	6.057 (0.202)		

k = 5, n = 8

16.000 (0.001)	15.900 (0.001)	13.500 (0.005)	13.400 (0.005)
12.300 (0.010)	12.200 (0.010)	10.600 (0.025)	10.500 (0.026)
9.200 (0.050)	9.100 (0.052)	7.700 (0.100)	7.600 (0.104)
6.100 (0.196)	6.000 (0.202)		

Table 39 CRITICAL VALUES FOR FRIEDMAN'S CHI SQUARE STATISTIC (Description, pp. 111-112)

k = 6, n = 2

10.000 (0.001) 9.714 (0.008) 9.429 (0.017) 9.143 (0.029)
8.857 (0.051) 8.286 (0.087) 8.000 (0.121) 7.429 (0.178)
7.143 (0.210)

k = 6, n = 3

13.286 (0.001) 13.095 (0.002) 12.524 (0.004) 12.333 (0.005)
11.762 (0.010) 11.571 (0.012) 10.810 (0.025) 10.619 (0.028)
9.857 (0.046) 9.667 (0.056) 8.714 (0.096) 8.524 (0.112)
7.571 (0.177) 7.381 (0.201)

k = 6, n = 4

15.286 (0.001) 15.143 (0.001) 13.571 (0.005) 13.429 (0.006)
12.714 (0.010) 12.571 (0.011) 11.429 (0.024) 11.286 (0.026)
10.286 (0.047) 10.143 (0.052) 9.000 (0.094) 8.857 (0.102)
7.429 (0.194) 7.286 (0.204)

k = 6, n = 5

16.429 (0.001) 16.314 (0.001) 14.257 (0.005) 14.143 (0.005)
13.229 (0.010) 13.114 (0.011) 11.743 (0.024) 11.629 (0.026)
10.486 (0.048) 10.371 (0.051) 9.000 (0.099) 8.886 (0.103)
7.400 (0.195) 7.286 (0.200)

k = 6, n = 6

17.048 (0.001) 16.952 (0.001) 14.762 (0.005) 14.667 (0.005)
13.619 (0.010) 13.524 (0.010) 12.000 (0.024) 11.905 (0.025)
10.571 (0.049) 10.476 (0.052) 9.048 (0.099) 8.952 (0.103)
7.333 (0.198) 7.238 (0.207)

Table 40 CRITICAL VALUES FOR PAGE'S L STATISTIC (Description: pp. 111-112)

k	n	0.200	0.100	0.050	0.025	0.010	0.005	0.001
		significance level, α						
4	3	80	83	84	86	87	88	89
	4	106	109	111	112	114	115	117
	5	132	134	137	139	141	142	145
	6	157	160	163	165	167	169	172
	7	183	186	189	191	193	195	198
	8	208	212	214	217	220	222	225
	9	233	237	240	243	246	248	252
	10	259	263	266	269	272	274	278
5	3	144	147	150	153	155	157	160
	4	190	194	197	200	204	206	210
	5	236	240	244	248	251	254	259
	6	281	287	291	295	299	302	307
	7	327	333	338	342	346	349	355
	8	373	379	384	389	393	397	403
	9	419	425	431	435	441	444	451
	10	464	471	477	482	487	491	499
6	3	233	239	244	248	252	255	260
	4	308	315	321	325	331	334	341
	5	383	391	397	403	409	412	420
	6	458	467	474	479	486	490	499
	7	533	542	550	556	563	568	577
	8	608	618	625	632	640	645	656
	9	682	693	701	708	717	722	733
	10	757	768	777	784	793	799	811
7	3	354	363	370	375	382	386	394
	4	469	479	487	494	501	506	516
	5	583	594	603	611	620	625	637
	6	697	709	719	728	737	744	757
	7	811	824	835	844	855	862	876
	8	924	939	950	960	972	979	995
	9	1038	1053	1065	1076	1088	1096	1113
	10	1152	1168	1181	1192	1205	1213	1231
8	3	511	523	532	540	550	556	567
	4	676	690	701	711	722	729	743
	5	841	857	869	880	893	901	917
	6	1006	1023	1037	1049	1063	1072	1090
	7	1171	1189	1204	1217	1232	1242	1262
	8	1335	1355	1371	1385	1401	1411	1433
	9	1499	1520	1537	1552	1569	1580	1603
	10	1663	1686	1704	1719	1737	1749	1774

Table 41 MULTIPLE COMPARISONS BASED ON FRIEDMAN RANK SUMS (Description: pp. 111-112)

k = 2

r	n = 2	n = 3	n = 4	n = 5	n = 6	n = 7	n = 8
8							0.008
7						0.016	
6					0.031		0.070
5				0.062		0.125	
4			0.125		0.219		0.289
3		0.250		0.375		0.453	
2	0.500		0.625		0.687		0.727

k = 3

r	n = 2	n = 3	n = 4	n = 5	n = 6	n = 7	n = 8
13							0.002
12						0.002	0.007
11					0.002	0.008	0.018
10				0.001	0.009	0.023	0.040
9				0.008	0.029	0.051	0.079
8			0.005	0.039	0.072	0.112	0.149
7			0.042	0.093	0.142	0.192	0.236
6		0.028	0.125	0.182	0.252	0.305	0.355
5		0.194	0.273	0.367	0.430	0.486	0.531
4	0.167	0.361	0.431	0.522	0.570	0.620	0.654
3	0.500	0.528	0.653	0.691	0.740	0.768	0.794

k = 4

r	n = 2	n = 3	n = 4	n = 5	n = 6	n = 7	n = 8
18							0.001
17						0.001	0.004
16						0.003	0.009
15					0.002	0.008	0.019
14				0.001	0.006	0.019	0.037
13				0.003	0.017	0.039	0.066
12			0.001	0.013	0.039	0.073	0.111
11			0.005	0.037	0.078	0.126	0.174
10			0.026	0.082	0.141	0.201	0.258
9		0.007	0.078	0.155	0.230	0.299	0.361
8		0.049	0.168	0.261	0.347	0.418	0.479
7		0.174	0.293	0.402	0.480	0.551	0.601
6	0.083	0.337	0.457	0.554	0.629	0.679	0.726
5	0.333	0.524	0.633	0.709	0.761	0.800	0.828
4	0.625	0.701	0.800	0.839	0.874	0.894	0.912
3	0.792	0.910	0.900	0.944	0.940	0.963	0.959

Table 41 MULTIPLE COMPARISONS BASED ON FRIEDMAN RANK SUMS (Description: pp. 111-112)

k = 5

r	n = 2	n = 3	n = 4	n = 5	n = 6	n = 7	n = 8
23							0.001
22						0.001	0.003
21						0.002	0.006
20					0.001	0.004	0.011
19					0.002	0.008	0.021
18					0.005	0.017	0.036
17				0.002	0.012	0.032	0.060
16				0.006	0.026	0.056	0.094
15			0.001	0.017	0.050	0.093	0.142
14			0.006	0.038	0.088	0.144	0.203
13			0.020	0.076	0.143	0.213	0.280
12		0.002	0.054	0.135	0.219	0.298	0.370
11		0.017	0.116	0.219	0.314	0.398	0.470
10		0.067	0.209	0.326	0.426	0.508	0.575
9		0.175	0.329	0.452	0.547	0.620	0.678
8	0.050	0.322	0.471	0.586	0.667	0.727	0.772
7	0.217	0.489	0.622	0.715	0.777	0.821	0.853
6	0.475	0.655	0.762	0.825	0.866	0.894	0.915
5	0.708	0.798	0.873	0.908	0.931	0.947	0.957

k = 6

r	n = 2	n = 3	n = 4	n = 5	n = 6
24					0.001
23					0.002
22					0.005
21				0.001	0.010
20				0.004	0.019
19				0.009	0.036
18			0.002	0.021	0.061
17			0.006	0.041	0.098
16			0.018	0.075	0.149
15		0.001	0.043	0.125	0.216
14		0.008	0.088	0.194	0.299
13		0.030	0.157	0.282	0.394
12		0.084	0.250	0.386	0.498
11		0.184	0.364	0.501	0.604
10	0.033	0.318	0.492	0.619	0.706
9	0.150	0.468	0.624	0.729	0.797
8	0.358	0.618	0.747	0.823	0.871
7	0.592	0.754	0.848	0.897	0.926
6	0.776	0.862	0.921	0.947	0.963
5	0.885	0.942	0.966	0.978	0.985

KOLMOGOROV-SMIRNOV TESTS (Tables 42, 43, 44 and 45)

Consider one- and two-sample problems. Let $X_1, \ldots, X_n$ be a random sample from a continuous distribution function F and $Y_1, \ldots, Y_m$ be an independent sample from continuous distribution function G. The empirical distribution function for X is defined to be $F_n(x)$ = (number of $X_i \leq x)/n$. Similarly, $G_m(y)$ is the empirical distribution for Y. The Kolmogorov-Smirnov statistics for specified F_o are

$$D = \sup_x |F_n(x) - F_o(x)|$$

$$D^+ = \sup_x \{F_n(x) - F_o(x)\}$$

$$K = \sup_x |F_n(x) - G_m(x)|$$

and

$$K^+ = \sup_x \{F_n(x) - G_m(x)\}$$

Table 42 CRITICAL VALUES FOR THE ONE-SAMPLE KOLMOGOROV-SMIRNOV STATISTIC (Table: p. 122)

One- and two-sided statistics to test H: $F = F_o$ are D^+ and D, respectively. The probability distributions of D^+ and D do not depend on F (when $F = F_o$) and are related by the approximate equality $P(D \geq s) \cong 2P(D^+ \geq s)$. Table 42 contains values s, for given n and p, such that

$$P(D \geq s) \cong 1 - p$$

and

$$P(D^+ \geq s) = \frac{1 - p}{2}$$

Table 43 CRITICAL VALUES FOR THE ONE-SIDED TWO-SAMPLE KOLMOGOROV-SMIRNOV STATISTIC (Table: pp. 123-125)

The table contains critical values for a one-sided test of $F = G$ against the alternative $F(x) \geq G(x)$ with strict inequality for at least one value of x. Using the notation of the general description for this group of tables, the smallest possible value of mnK^+, say d, is tabulated such that when $F = G$, and m, n and γ are as given,

$$P(mnK^+ \geq d) \leq 1 - \gamma$$

Table 44 CRITICAL VALUES FOR THE TWO-SIDED TWO-SAMPLE KOLMOGOROV-SMIRNOV STATISTIC (Table: pp. 126-128)

The table contains critical values for a two-sided test of the hypothesis $F = G$. Using the notation of the general description for this group of tables, the smallest possible value of mnK, say d, is tabulated such that

$$P(mnK \geq d) \leq 1 - \gamma$$

Table 45 LIMITING CRITICAL VALUES OF THE KOLMOGOROV-SMIRNOV STATISTICS (Table: p. 129)

Using the notation of the general description for this group of tables, under $F = F_o$ or $F = G$,

$$\lim_{n\to\infty} P(\sqrt{n}D \leq z) = \lim_{\min(n,m)\to\infty} P\left(\sqrt{\frac{nm}{n+m}}\, K \leq z\right) = L(z)$$

Tabulated are values of z and $L(z)$.

Table 42 CRITICAL VALUES FOR THE ONE-SAMPLE KOLMOGOROV-SMIRNOV STATISTIC (Description: pp. 120-121)

n	p = 0.80	0.90	0.95	0.98	0.99
1	0.90000	0.95000	0.97500	0.99000	0.99500
2	0.68377	0.77639	0.84189	0.90000	0.92929
3	0.56481	0.63604	0.70760	0.78456	0.82900
4	0.49265	0.56522	0.62394	0.68887	0.73424
5	0.44698	0.50945	0.56328	0.62718	0.66853
6	0.41037	0.46799	0.51926	0.57741	0.61661
7	0.38148	0.43607	0.48342	0.53844	0.57581
8	0.35831	0.40962	0.45427	0.50654	0.54179
9	0.33910	0.38746	0.43001	0.47960	0.51332
10	0.32260	0.36866	0.40925	0.45662	0.48893
11	0.30829	0.35242	0.39122	0.43670	0.46770
12	0.29577	0.33815	0.37543	0.41918	0.44905
13	0.28470	0.32549	0.36143	0.40362	0.43247
14	0.27481	0.31417	0.34890	0.38969	0.41762
15	0.26589	0.30397	0.33760	0.37713	0.40420
16	0.25778	0.29472	0.32733	0.36571	0.39201
17	0.25039	0.28627	0.31796	0.35528	0.38086
18	0.24360	0.27851	0.30936	0.34569	0.37062
19	0.23735	0.27136	0.30143	0.33685	0.36117
20	0.23156	0.26473	0.29408	0.32866	0.35241
21	0.22617	0.25858	0.28724	0.32104	0.34426
22	0.22115	0.25283	0.28087	0.31394	0.33666
23	0.21646	0.24746	0.27490	0.30728	0.32954
24	0.21205	0.24242	0.26931	0.30104	0.32286
25	0.20790	0.23768	0.26404	0.29516	0.31657
26	0.20399	0.23320	0.25908	0.28962	0.31063
27	0.20030	0.22898	0.25438	0.28438	0.30502
28	0.19680	0.22497	0.24993	0.27942	0.29971
29	0.19348	0.22117	0.24571	0.27471	0.29466
30	0.19032	0.21756	0.24170	0.27023	0.28986
31	0.18732	0.21412	0.23788	0.26596	0.28529
32	0.18445	0.21085	0.23424	0.26189	0.28094
33	0.18171	0.20771	0.23076	0.25801	0.27677
34	0.17909	0.20472	0.22743	0.25429	0.27279
35	0.17659	0.20185	0.22425	0.25073	0.26897
36	0.17418	0.19910	0.22119	0.24732	0.26532
37	0.17188	0.19646	0.21826	0.24404	0.26180
38	0.16966	0.19392	0.21544	0.24089	0.25843
39	0.16753	0.19148	0.21273	0.23786	0.25518
40	0.16547	0.18913	0.21012	0.23494	0.25205
42	0.16158	0.18468	0.20517	0.22941	0.24613
44	0.15796	0.18053	0.20056	0.22426	0.24060
46	0.15457	0.17665	0.19625	0.21944	0.23544
48	0.15139	0.17301	0.19221	0.21493	0.23059
50	0.14840	0.16959	0.18841	0.21068	0.22604

Table 43 CRITICAL VALUES FOR THE ONE-SIDED TWO-SAMPLE KOLMOGOROV-SMIRNOV STATISTIC (Description: pp. 120-121)

(Description: pp. 120-121)

		Y									
m	n	0.70	0.75	0.80	0.85	0.90	0.95	0.975	0.99	0.995	0.999
4	4	12	12	12	12	16	16	16	-	-	-
	5	10	11	12	12	15	16	20	20	-	-
	6	12	14	14	14	16	18	20	24	24	-
	7	13	14	16	17	20	21	24	28	28	-
	8	16	20	20	20	24	24	28	32	32	-
4	9	18	18	19	20	23	27	28	32	36	-
	10	18	20	22	22	24	28	30	36	36	40
5	5	15	15	15	20	20	20	25	25	25	-
	6	13	14	15	18	19	24	24	30	30	-
	7	15	16	18	20	21	25	28	30	35	-
	8	16	19	20	22	24	27	30	35	35	40
	9	18	20	21	25	26	30	35	36	40	45
5	10	25	25	25	30	30	35	40	40	45	50
6	6	18	18	24	24	24	30	30	36	36	-
	7	17	18	21	22	24	28	30	35	36	42
	8	20	20	22	26	26	30	34	40	40	48
	9	21	24	27	27	30	33	39	42	45	54
	10	22	26	26	30	32	36	40	44	48	54
6	11	24	26	27	31	33	38	43	49	54	60
	12	30	30	36	36	42	48	48	54	60	66
7	7	21	28	28	28	35	35	42	42	42	49
	8	21	24	25	27	28	34	40	42	48	56
	9	22	26	27	29	33	36	42	47	49	56
	10	25	28	29	32	35	40	46	50	53	60
	11	27	30	31	34	38	44	48	55	59	66
7	12	29	30	34	36	41	46	53	58	60	70
	13	30	32	36	38	44	50	56	63	65	77
	14	35	42	42	49	49	56	63	70	77	84
8	8	32	32	32	32	40	40	48	48	56	64
	9	24	28	30	31	36	40	46	54	55	63
	10	28	30	32	34	40	44	48	56	60	70
	11	29	32	34	37	42	48	53	61	64	72
	12	36	36	40	44	48	52	60	64	68	76
8	13	33	36	39	43	48	54	62	67	72	83
	14	36	40	42	46	50	58	64	72	76	88
	15	37	41	44	49	52	60	67	75	81	90
	16	48	48	56	56	64	72	80	88	88	104

Table 43 CRITICAL VALUES FOR THE ONE SIDED TWO-SAMPLE KOLMOGOROV-SMIRNOV STATISTIC
(Description: pp. 120-121)

		γ									
m	n	0.70	0.75	0.80	0.85	0.90	0.95	0.975	0.99	0.995	0.999
9	9	36	36	36	45	45	54	54	63	63	72
	10	30	32	34	36	43	50	53	61	63	72
	11	32	35	37	41	45	52	59	63	70	79
	12	36	39	42	45	51	57	63	69	75	84
	13	37	39	43	46	51	59	65	73	78	90
9	14	39	42	45	49	54	63	70	80	84	94
	15	42	45	48	54	60	69	75	84	90	99
	16	44	47	51	55	62	69	78	87	94	108
	17	46	49	54	57	65	74	82	92	99	110
	18	54	63	63	72	72	81	90	99	108	126
10	10	40	40	50	50	50	60	70	70	80	90
	11	35	37	39	45	48	57	60	69	77	88
	12	38	40	42	48	52	60	66	74	80	90
	13	39	42	45	50	55	64	70	78	84	97
	14	42	46	50	54	60	68	74	84	90	102
10	15	50	50	55	60	65	75	80	90	100	110
	16	46	50	56	60	66	76	84	94	100	112
	17	49	52	58	62	69	79	89	99	106	119
	18	52	56	60	66	72	82	92	104	108	124
	19	53	57	63	66	74	85	94	104	113	131
10	20	60	70	70	80	90	100	110	120	130	140
11	11	44	44	55	55	66	66	77	88	88	99
	12	40	42	44	51	54	64	72	77	86	98
	13	42	45	49	54	60	67	75	86	91	104
	14	44	48	52	56	63	73	82	90	96	110
	15	47	51	55	60	66	76	84	95	102	117
11	16	50	53	57	63	69	80	89	100	106	121
	17	52	57	60	66	74	85	93	104	110	131
	18	54	59	64	70	77	88	97	108	118	133
	19	57	62	66	73	80	92	102	114	122	141
	20	59	65	69	76	85	96	107	118	127	147
12	12	48	60	60	60	72	72	84	96	96	108
	13	45	47	53	57	65	71	81	92	95	108
	14	48	50	56	62	68	78	86	94	104	118
	15	51	57	60	66	72	84	93	102	108	123
	16	56	60	64	72	76	88	96	108	116	132
12	17	55	60	66	71	78	90	100	112	119	136
	18	60	66	72	78	84	96	108	120	126	144
	19	60	65	71	77	85	99	108	121	130	149
	20	64	72	76	84	92	104	116	128	140	156

Table 43 CRITICAL VALUES FOR THE ONE-SIDED TWO-SAMPLE KOLMOGOROV-SMIRNOV STATISTIC (Description: pp. 120-121)

m	n	γ 0.70	0.75	0.80	0.85	0.90	0.95	0.975	0.99	0.995	0.999
13	13	65	65	65	78	78	91	91	104	117	130
	14	50	52	59	63	72	78	89	102	104	126
	15	52	57	62	68	75	87	96	107	115	130
	16	56	60	65	72	79	91	101	112	121	137
	17	58	63	68	75	83	96	105	118	127	143
13	18	61	66	72	78	87	99	110	123	131	151
	19	63	68	75	81	91	104	114	130	138	157
	20	67	71	78	84	95	108	120	135	143	162
14	14	70	70	70	84	84	98	112	112	126	140
	15	55	61	66	69	80	92	98	111	123	138
	16	58	66	70	76	84	96	106	120	126	144
	17	61	67	72	78	88	100	111	125	134	151
	18	64	70	76	82	92	104	116	130	140	160
14	19	68	73	79	86	95	110	121	135	148	167
	20	72	76	82	90	100	114	126	142	152	172
15	15	75	75	75	90	90	105	120	135	135	150
	16	60	68	72	80	87	101	114	120	133	149
	17	65	71	75	84	91	105	116	131	142	161
	18	69	75	81	87	99	111	123	138	147	168
	19	71	76	82	90	100	114	127	142	152	175
15	20	80	85	90	95	110	125	135	150	160	185
16	16	80	80	96	96	112	112	128	144	160	176
	17	69	74	78	88	94	109	124	139	143	170
	18	72	78	82	92	100	116	128	142	154	174
	19	75	80	87	95	104	120	133	151	160	183
	20	80	88	92	100	112	128	140	156	168	192
17	17	85	85	102	102	119	136	136	153	170	187
	18	75	80	85	96	102	118	133	150	164	184
	19	78	83	90	99	109	126	141	158	166	190
	20	81	87	95	103	113	130	146	163	175	198
18	18	90	108	108	108	126	144	162	180	180	198
	19	81	86	96	103	116	133	142	160	176	196
	20	86	90	100	108	120	136	152	170	182	208
19	19	95	114	114	133	133	152	171	190	190	228
	20	88	93	104	110	125	144	160	171	187	209
20	20	100	120	120	140	140	160	180	200	220	240

Table 44 CRITICAL VALUES FOR THE TWO-SIDED TWO-SAMPLE KOLMOGOROV-SMIRNOV STATISTIC
(Description: pp. 120-121)

m	n	γ 0.70	0.75	0.80	0.85	0.90	0.95	0.975	0.99	0.995	0.999
4	4	12	12	16	16	16	16	-	-	-	-
	5	12	15	15	15	16	20	20	-	-	-
	6	14	16	16	18	18	20	24	24	-	-
	7	17	17	20	20	21	24	28	28	-	-
	8	20	20	24	24	24	28	28	32	32	-
4	9	20	23	23	24	27	28	32	36	36	-
	10	22	24	24	26	28	30	36	36	40	-
5	5	20	20	20	20	20	25	25	25	-	-
	6	18	18	19	20	24	24	30	30	30	-
	7	20	20	21	23	25	28	30	35	35	-
	8	22	22	24	25	27	30	32	35	40	-
	9	25	25	26	27	30	35	36	40	45	45
5	10	30	30	30	35	35	40	40	45	45	50
6	6	24	24	24	24	30	30	36	36	36	-
	7	22	23	24	24	28	30	35	36	42	-
	8	26	26	26	28	30	34	36	40	42	48
	9	27	30	30	33	33	39	42	45	48	54
	10	30	30	32	34	36	40	44	48	50	60
6	11	31	32	33	37	38	43	48	54	55	66
	12	36	42	42	42	48	48	54	60	60	66
7	7	28	28	35	35	35	42	42	42	49	49
	8	27	28	28	32	34	40	41	48	48	56
	9	29	31	33	35	36	42	45	49	54	63
	10	32	33	35	39	40	46	49	53	56	63
	11	34	35	38	41	44	48	52	59	63	70
7	12	36	39	41	44	46	53	56	60	65	72
	13	38	42	44	45	50	56	58	65	70	78
	14	49	49	49	56	56	63	70	77	77	84
8	8	32	40	40	40	40	48	48	56	56	64
	9	31	32	36	38	40	46	48	55	56	64
	10	34	36	40	42	44	48	54	60	62	70
	11	37	39	42	44	48	53	58	64	66	77
	12	44	44	48	48	52	60	64	68	72	80
8	13	43	44	48	51	54	62	65	72	78	88
	14	46	48	50	54	58	64	70	76	82	90
	15	49	50	52	57	60	67	74	81	88	97
	16	56	56	64	64	72	80	80	88	96	104

Table 44 CRITICAL VALUES FOR THE TWO-SIDED TWO-SAMPLE KOLMOGOROV-SMIRNOV STATISTIC
(Description: pp. 120-121)

		γ									
m	n	0.70	0.75	0.80	0.85	0.90	0.95	0.975	0.99	0.995	0.999
9	9	45	45	45	45	54	54	63	63	72	72
	10	36	41	43	44	50	53	60	63	70	80
	11	41	43	45	48	52	59	63	70	72	81
	12	45	48	51	54	57	63	69	75	78	87
	13	46	50	51	55	59	65	72	78	82	91
9	14	49	52	54	58	63	70	76	84	89	98
	15	54	57	60	63	69	75	81	90	93	105
	16	55	58	62	64	69	78	85	94	99	110
	17	57	59	65	67	74	82	90	99	102	117
	18	72	72	72	81	81	90	99	108	117	126
10	10	50	50	50	60	60	70	70	80	80	90
	11	45	47	48	50	57	60	68	77	79	89
	12	48	50	52	56	60	66	72	80	84	96
	13	50	52	55	58	64	70	77	84	90	100
	14	54	56	60	62	68	74	82	90	96	106
10	15	60	65	65	70	75	80	90	100	105	115
	16	60	62	66	70	76	84	90	100	108	118
	17	62	65	69	73	79	89	96	106	110	126
	18	66	68	72	76	82	92	100	108	116	132
	19	66	72	74	81	85	94	103	113	122	133
10	20	80	80	90	90	100	110	120	130	130	150
11	11	55	55	66	66	66	77	77	88	88	99
	12	51	53	54	60	64	72	76	86	88	99
	13	54	56	60	64	67	75	84	91	97	108
	14	56	60	63	66	73	82	87	96	101	115
	15	60	64	66	72	76	84	94	102	109	120
11	16	63	67	69	74	80	89	96	106	112	127
	17	66	70	74	77	85	93	102	110	119	132
	18	70	72	77	82	88	97	107	118	125	140
	19	73	76	80	86	92	102	111	122	130	146
	20	76	79	85	89	96	107	116	127	136	154
12	12	60	72	72	72	72	84	96	96	108	120
	13	57	59	65	68	71	81	84	95	104	117
	14	62	64	68	72	78	86	94	104	108	120
	15	66	69	72	78	84	93	99	108	117	129
	16	68	72	76	80	88	96	104	116	124	136
12	17	71	74	78	83	90	100	108	119	127	141
	18	78	84	84	90	96	108	120	126	138	150
	19	77	82	85	90	99	108	120	130	140	156
	20	84	88	92	96	104	116	124	140	148	164

Table 44 CRITICAL VALUES FOR THE TWO SIDED TWO-SAMPLE KOLMOGOROV-SMIRNOV STATISTIC
(Description: pp. 120-121)

		γ									
m	n	0.70	0.75	0.80	0.85	0.90	0.95	0.975	0.99	0.995	0.999
13	13	65	78	78	78	91	91	104	117	117	130
	14	63	65	72	75	78	89	100	104	115	129
	15	68	72	75	79	87	96	104	115	122	137
	16	72	75	79	83	91	101	111	121	128	143
	17	75	78	83	88	96	105	114	127	135	152
13	18	78	82	87	92	99	110	120	131	141	156
	19	81	86	91	95	104	114	126	138	145	164
	20	84	89	95	100	108	120	130	143	154	169
14	14	84	84	84	98	98	112	112	126	126	154
	15	69	76	80	82	92	98	110	123	125	140
	16	76	80	84	88	96	106	116	126	136	152
	17	78	83	88	92	100	111	122	134	140	159
	18	82	86	92	98	104	116	126	140	148	166
14	19	86	91	95	101	110	121	133	148	154	176
	20	90	94	100	106	114	126	138	152	160	180
15	15	90	90	90	105	105	120	135	135	150	165
	16	80	84	87	90	101	114	119	133	144	162
	17	84	87	91	97	105	116	129	142	148	165
	18	87	93	99	102	111	123	135	147	156	174
	19	90	95	100	105	114	127	141	152	161	180
15	20	95	105	110	115	125	135	150	160	170	195
16	16	96	96	112	112	112	128	144	160	160	176
	17	88	91	94	104	109	124	136	143	157	174
	18	92	96	100	108	116	128	140	154	162	186
	19	95	100	104	112	120	133	145	160	170	190
	20	100	108	112	120	128	140	156	168	180	200
17	17	102	102	119	119	136	136	153	170	170	204
	18	96	99	102	113	118	133	148	164	168	187
	19	99	103	109	116	126	141	151	166	179	200
	20	102	109	113	121	130	146	160	175	186	209
18	18	108	126	126	126	144	162	162	180	198	216
	19	103	106	116	121	133	142	159	176	180	212
	20	108	112	120	126	136	152	166	182	194	214
19	19	133	133	133	152	152	171	190	190	209	228
	20	110	114	125	130	144	160	169	187	204	225
20	20	140	140	140	160	160	180	200	220	220	260

Table 45 LIMITING CRITICAL VALUES OF THE KOLMOGOROV-SMIRNOV STATISTICS
(Description: pp. 120-121)

z	L(z)	z	L(z)	z	L(z)
0.34	0.0002	0.79	0.4395	1.24	0.9076
0.35	0.0003	0.80	0.4559	1.25	0.9121
0.36	0.0005	0.81	0.4720	1.26	0.9164
0.37	0.0008	0.82	0.4880	1.27	0.9206
0.38	0.0013	0.83	0.5038	1.28	0.9245
0.39	0.0019	0.84	0.5194	1.29	0.9283
0.40	0.0028	0.85	0.5347	1.30	0.9319
0.41	0.0040	0.86	0.5497	1.32	0.9387
0.42	0.0055	0.87	0.5645	1.34	0.9449
0.43	0.0074	0.88	0.5791	1.36	0.9505
0.44	0.0097	0.89	0.5933	1.38	0.9557
0.45	0.0126	0.90	0.6073	1.40	0.9603
0.46	0.0160	0.91	0.6209	1.42	0.9646
0.47	0.0200	0.92	0.6343	1.44	0.9684
0.48	0.0247	0.93	0.6473	1.46	0.9718
0.49	0.0300	0.94	0.6601	1.48	0.9750
0.50	0.0361	0.95	0.6725	1.50	0.9778
0.51	0.0428	0.96	0.6846	1.52	0.9803
0.52	0.0503	0.97	0.6964	1.54	0.9826
0.53	0.0585	0.98	0.7079	1.56	0.9846
0.54	0.0675	0.99	0.7191	1.58	0.9864
0.55	0.0772	1.00	0.7300	1.60	0.9880
0.56	0.0876	1.01	0.7406	1.62	0.9895
0.57	0.0987	1.02	0.7508	1.64	0.9908
0.58	0.1104	1.03	0.7608	1.66	0.9919
0.59	0.1228	1.04	0.7704	1.68	0.9929
0.60	0.1357	1.05	0.7798	1.70	0.9938
0.61	0.1492	1.06	0.7889	1.72	0.9946
0.62	0.1633	1.07	0.7976	1.74	0.9953
0.63	0.1778	1.08	0.8061	1.76	0.9959
0.64	0.1927	1.09	0.8143	1.78	0.9965
0.65	0.2080	1.10	0.8223	1.80	0.9969
0.66	0.2236	1.11	0.8300	1.82	0.9973
0.67	0.2396	1.12	0.8374	1.84	0.9977
0.68	0.2558	1.13	0.8445	1.86	0.9980
0.69	0.2722	1.14	0.8514	1.88	0.9983
0.70	0.2888	1.15	0.8580	1.90	0.9985
0.71	0.3055	1.16	0.8644	1.92	0.9987
0.72	0.3223	1.17	0.8706	1.94	0.9989
0.73	0.3391	1.18	0.8765	1.96	0.9991
0.74	0.3560	1.19	0.8823	1.98	0.9992
0.75	0.3728	1.20	0.8878	2.00	0.9993
0.76	0.3896	1.21	0.8930	2.02	0.9994
0.77	0.4064	1.22	0.8981	2.04	0.9995
0.78	0.4230	1.23	0.9030	2.06	0.9996

Table 46 CRITICAL VALUES FOR FISHER'S EXACT TEST FOR TWO BY TWO TABLES (Table: pp. 132-145)

To use this table take the 2×2 table under consideration and, by interchanging rows and/or columns, arrange it so that (i) $A \geq B$ and (ii) $a/A \geq b/B$, or equivalently $aB \geq bA$, where the table entries and totals are

		Total
a	A - a	A
b	B - b	B

Find the table entry corresponding to A, B and a. (The entries are ordered lexicographically by A, B and a where $A \geq B$ and $A \geq a$.) The two columns under $\alpha = 0.05$ labeled b and P give (1) the nominal 0.05 critical value for a one-sided test of the true proportion corresponding to a/A being greater than the true proportion corresponding to b/B and (2) the P-value corresponding to the critical value (usually less than 0.05 due to the discreteness of the distribution). The hypothesis of equality of the true proportions is rejected if the observed b is less than or equal to the entry under the column labeled b. A "-" in a column indicates that no such critical value exists for the (A, B, a) triple. For a two-sided test the P-value should be multiplied by two. The next two columns give the nominal 0.01 critical value and P-value.

Example: Perform a two-sided test at $\alpha = 0.10$ for the table:

Yes	No	
8	2	Group 1
3	12	Group 2

Note that Group 2 has a higher proportion of no answers.

Rearrange the table to:

No	Yes	Total	
12	3	15	Group 2
2	8	10	Group 1

The critical value is 3, $P = 0.018 \times 2$. Thus, reject equality of proportions at the $\alpha = 0.05$ level. The exact P-value in this case is 0.005×2 (from the $\alpha = 0.01$ column).

Table 46 CRITICAL VALUES FOR FISHER'S EXACT TEST FOR TWO BY TWO TABLES
(Description: pp. 130-131)

			$\alpha \le 0.05$		$\alpha \le 0.01$	
A	B	a	b	P	b	P
3	3	3	0	.050	-	
4	4	4	0	.014	-	
	3	4	0	.029	-	
5	5	5	1	.024	0	.004
		4	0	.024	-	
	4	5	1	.048	0	.008
		4	0	.040	-	
	3	5	0	.018	-	
	2	5	0	.048	-	
6	6	6	2	.030	1	.008
		5	1	.040	0	.008
		4	0	.030	-	
	5	6	1	.015	0	.002
		5	0	.013	-	
		4	0	.045	-	
	4	6	1	.033	0	.005
		5	0	.024	-	
	3	6	0	.012	-	
		5	0	.048	-	
	2	6	0	.036	-	
7	7	7	3	.035	1	.002
		6	1	.015	0	.002
		5	0	.010	-	
		4	0	.035	-	
	6	7	2	.021	1	.005
		6	1	.025	0	.004
		5	0	.016	-	
		4	0	.049	-	
7	5	7	2	.045	0	.001
		6	1	.045	0	.008
		5	0	.027	-	
	4	7	1	.024	0	.003
		6	0	.015	-	
		5	0	.045	-	
	3	7	0	.008	0	.008
		6	0	.033	-	
	2	7	0	.028	-	
8	8	8	4	.038	2	.003
		7	2	.020	1	.005
		6	1	.020	0	.003
		5	0	.013	-	
		4	0	.038	-	
	7	8	3	.026	2	.007
		7	2	.035	1	.009
		6	1	.032	0	.006
		5	0	.019	-	
	6	8	2	.015	1	.003
		7	1	.016	0	.002
		6	0	.009	0	.009
		5	0	.028	-	
	5	8	2	.035	1	.007
		7	1	.032	0	.005
		6	0	.016	-	
		5	0	.044	-	
	4	8	1	.018	0	.002
		7	0	.010	-	
		6	0	.030	-	
	3	8	0	.006	0	.006
		7	0	.024	-	
	2	8	0	.022	-	

Table 46 CRITICAL VALUES FOR FISHER'S EXACT TEST FOR TWO BY TWO TABLES
(Description: pp. 130-131)

			α ≤ 0.05		α ≤ 0.01	
A	B	a	b	P	b	P
9	9	9	5	.041	3	.005
		8	3	.025	2	.008
		7	2	.028	1	.008
		6	1	.025	0	.005
		5	0	.015	–	
		4	0	.041	–	
	8	9	4	.029	3	.009
		8	3	.043	1	.003
		7	2	.044	0	.002
		6	1	.036	0	.007
		5	0	.020	–	
	7	9	3	.019	2	.005
		8	2	.024	1	.006
		7	1	.020	0	.003
		6	0	.010	–	
		5	0	.029	–	
	6	9	3	.044	1	.002
		8	2	.047	0	.001
		7	1	.035	0	.006
		6	0	.017	–	
		5	0	.042	–	
	5	9	2	.027	1	.005
		8	1	.023	0	.003
		7	0	.010	–	
		6	0	.028	–	
	4	9	1	.014	0	.001
		8	0	.007	0	.007
		7	0	.021	–	
		6	0	.049	–	
	3	9	1	.045	0	.005
		8	0	.018	–	
		7	0	.045	–	
	2	9	0	.018	–	

			α ≤ 0.05		α ≤ 0.01	
A	B	a	b	P	b	P
10	10	10	6	.043	4	.005
		9	4	.029	3	.010
		8	3	.035	1	.003
		7	2	.035	1	.010
		6	1	.029	0	.005
		5	0	.016	–	
		4	0	.043	–	
	9	10	5	.033	3	.003
		9	4	.050	2	.005
		8	2	.019	1	.004
		7	1	.015	0	.002
		6	1	.040	0	.008
		5	0	.022	–	
	8	10	4	.023	3	.007
		9	3	.032	2	.009
		8	2	.031	1	.008
		7	1	.023	0	.004
		6	0	.011	–	
		5	0	.029	–	
	7	10	3	.015	2	.003
		9	2	.018	1	.004
		8	1	.013	0	.002
		7	1	.036	0	.006
		6	0	.017	–	
		5	0	.041	–	
	6	10	3	.036	2	.008
		9	2	.036	1	.008
		8	1	.024	0	.003
		7	0	.010	–	
		6	0	.026	–	
	5	10	2	.022	1	.004
		9	1	.017	0	.002
		8	1	.047	0	.007
		7	0	.019	–	
		6	0	.042	–	
	4	10	1	.011	0	.001
		9	1	.041	0	.005
		8	0	.015	–	
		7	0	.035	–	
	3	10	1	.038	0	.003
		9	0	.014	–	
		8	0	.035	–	
	2	10	0	.015	–	
		9	0	.045	–	

Table 46 CRITICAL VALUES FOR FISHER'S EXACT TEST FOR TWO BY TWO TABLES
(Description: pp. 130-131)

A	B	a	$\alpha \le 0.05$ b	$\alpha \le 0.05$ P	$\alpha \le 0.01$ b	$\alpha \le 0.01$ P
11	11	11	7	.045	5	.006
		10	5	.032	3	.004
		9	4	.040	2	.004
		8	3	.043	1	.004
		7	2	.040	0	.002
		6	1	.032	0	.006
		5	0	.018	-	
		4	0	.045	-	
	10	11	6	.035	4	.004
		10	4	.021	3	.007
		9	3	.024	2	.007
		8	2	.023	1	.006
		7	1	.017	0	.003
		6	1	.043	0	.009
		5	0	.023	-	
	9	11	5	.026	4	.008
		10	4	.038	2	.003
		9	3	.040	1	.003
		8	2	.035	1	.009
		7	1	.025	0	.004
		6	0	.012	-	
		5	0	.030	-	
	8	11	4	.018	3	.005
		10	3	.024	2	.006
		9	2	.022	1	.005
		8	1	.015	0	.002
		7	1	.037	0	.007
		6	0	.017	-	
		5	0	.040	-	
	7	11	4	.043	2	.002
		10	3	.047	1	.002
		9	2	.039	1	.009
		8	1	.025	0	.004
		7	0	.010	-	
		6	0	.025	-	
	6	11	3	.029	2	.006
		10	2	.028	1	.005
		9	1	.018	0	.002
		8	1	.043	0	.007
		7	0	.017	-	
		6	0	.037	-	
	5	11	2	.018	1	.003
		10	1	.013	0	.001
		9	1	.036	0	.005
		8	0	.013	-	
		7	0	.029	-	
11	4	11	1	.009	1	.009
		10	1	.033	0	.004
		9	0	.011	-	
		8	0	.026	-	
	3	11	1	.033	0	.003
		10	0	.011	-	
		9	0	.027	-	
	2	11	0	.013	-	
		10	0	.038	-	
12	12	12	8	.047	6	.007
		11	6	.034	4	.005
		10	5	.045	3	.006
		9	4	.050	2	.006
		8	3	.050	1	.005
		7	2	.045	0	.002
		6	1	.034	0	.007
		5	0	.019	-	
		4	0	.047	-	
	11	12	7	.037	5	.005
		11	5	.024	4	.008
		10	4	.029	2	.003
		9	3	.030	2	.009
		8	2	.026	1	.007
		7	1	.019	0	.003
		6	1	.045	0	.009
		5	0	.024	-	
	10	12	6	.029	5	.010
		11	5	.043	3	.005
		10	4	.048	2	.005
		9	3	.046	1	.004
		8	2	.038	0	.002
		7	1	.026	0	.005
		6	0	.012	-	
		5	0	.030	-	
	9	12	5	.021	4	.006
		11	4	.029	3	.009
		10	3	.029	2	.008
		9	2	.024	1	.006
		8	1	.016	0	.002
		7	1	.037	0	.007
		6	0	.017	-	
		5	0	.039	-	
	8	12	5	.049	3	.004
		11	3	.018	2	.004
		10	2	.015	1	.003

Table 46 CRITICAL VALUES FOR FISHER'S EXACT TEST FOR TWO BY TWO TABLES
(Description: pp. 130-131)

			$\alpha \leq 0.05$		$\alpha \leq 0.01$	
A	B	a	b	P	b	P
12	8	9	2	.040	1	.010
		8	1	.025	0	.004
		7	0	.010	–	
		6	0	.024	–	
	7	12	4	.036	3	.009
		11	3	.038	2	.010
		10	2	.029	1	.006
		9	1	.017	0	.002
		8	1	.040	0	.007
		7	0	.016	–	
		6	0	.034	–	
	6	12	3	.025	2	.005
		11	2	.022	1	.004
		10	1	.013	0	.002
		9	1	.032	0	.005
		8	0	.011	–	
		7	0	.025	–	
		6	0	.050	–	
	5	12	2	.015	1	.002
		11	1	.010	1	.010
		10	1	.028	0	.003
		9	0	.009	0	.009
		8	0	.020	–	
		7	0	.041	–	
	4	12	1	.007	1	.007
		11	1	.027	0	.003
		10	0	.008	0	.008
		9	0	.019	–	
		8	0	.038	–	
	3	12	1	.029	0	.002
		11	0	.009	0	.009
		10	0	.022	–	
		9	0	.044	–	
	2	12	0	.011	–	
		11	0	.033	–	
13	13	13	9	.048	7	.007
		12	7	.037	5	.006
		11	6	.048	4	.008
		10	4	.024	3	.008
		9	3	.024	2	.008
		8	2	.021	1	.006
		7	2	.048	0	.003
		6	1	.037	0	.007
		5	0	.020	–	
		4	0	.048	–	

			$\alpha \leq 0.05$		$\alpha \leq 0.01$	
A	B	a	b	P	b	P
13	12	13	8	.039	6	.005
		12	6	.027	5	.010
		11	5	.033	3	.004
		10	4	.036	2	.004
		9	3	.034	1	.003
		8	2	.029	1	.008
		7	1	.020	0	.004
		6	1	.046	0	.010
		5	0	.024	–	
	11	13	7	.031	5	.003
		12	6	.048	4	.006
		11	4	.021	3	.007
		10	3	.021	2	.006
		9	3	.050	1	.004
		8	2	.040	0	.002
		7	1	.027	0	.005
		6	0	.013	–	
		5	0	.030	–	
	10	13	6	.024	5	.007
		12	5	.035	3	.003
		11	4	.037	2	.003
		10	3	.033	1	.002
		9	2	.026	1	.006
		8	1	.017	0	.003
		7	1	.038	0	.007
		6	0	.017	–	
		5	0	.038	–	
	9	13	5	.017	4	.005
		12	4	.023	3	.007
		11	3	.022	2	.006
		10	2	.017	1	.004
		9	2	.040	0	.001
		8	1	.025	0	.004
		7	0	.010	–	
		6	0	.023	–	
		5	0	.049	–	
	8	13	5	.042	3	.003
		12	4	.047	2	.003
		11	3	.041	1	.002
		10	2	.029	1	.007
		9	1	.017	0	.002
		8	1	.037	0	.006
		7	0	.015	–	
		6	0	.032	–	
	7	13	4	.031	3	.007
		12	3	.031	2	.007
		11	2	.022	1	.004
		10	1	.012	0	.002

Table 46 CRITICAL VALUES FOR FISHER'S EXACT TEST FOR TWO BY TWO TABLES
(Description: pp. 130-131)

A	B	a	$\alpha \leq 0.05$		$\alpha \leq 0.01$	
			b	P	b	P
13	7	9	1	.029	0	.004
		8	0	.010	-	
		7	0	.022	-	
		6	0	.044	-	
	6	13	3	.021	2	.004
		12	2	.017	1	.003
		11	2	.046	1	.010
		10	1	.024	0	.003
		9	1	.050	0	.008
		8	0	.017	-	
		7	0	.034	-	
	5	13	2	.012	1	.002
		12	2	.044	1	.008
		11	1	.022	0	.002
		10	1	.047	0	.007
		9	0	.015	-	
		8	0	.029	-	
	4	13	2	.044	1	.006
		12	1	.022	0	.002
		11	0	.006	0	.006
		10	0	.015	-	
		9	0	.029	-	
	3	13	1	.025	0	.002
		12	0	.007	0	.007
		11	0	.018	-	
		10	0	.036	-	
	2	13	0	.010	0	.010
		12	0	.029	-	
14	14	14	10	.049	8	.008
		13	8	.038	6	.006
		12	6	.023	5	.009
		11	5	.027	3	.004
		10	4	.028	2	.003
		9	3	.027	2	.009
		8	2	.023	1	.006
		7	1	.016	0	.003
		6	1	.038	0	.008
		5	0	.020	-	
		4	0	.049	-	
	13	14	9	.041	7	.006
		13	7	.029	5	.004
		12	6	.037	4	.005
		11	5	.041	3	.006
		10	4	.041	2	.005

A	B	a	$\alpha \leq 0.05$		$\alpha \leq 0.01$	
			b	P	b	P
14	13	9	3	.038	1	.003
		8	2	.031	1	.009
		7	1	.021	0	.004
		6	1	.048	-	
		5	0	.025	-	
	12	14	8	.033	6	.004
		13	6	.021	5	.007
		12	5	.025	4	.009
		11	4	.026	3	.009
		10	3	.024	2	.007
		9	2	.019	1	.005
		8	2	.042	0	.002
		7	1	.028	0	.005
		6	0	.013	-	
		5	0	.030	-	
	11	14	7	.026	6	.009
		13	6	.039	4	.004
		12	5	.043	3	.005
		11	4	.042	2	.004
		10	3	.036	1	.003
		9	2	.027	1	.007
		8	1	.017	0	.003
		7	1	.038	0	.007
		6	0	.017	-	
		5	0	.038	-	
	10	14	6	.020	5	.006
		13	5	.028	4	.009
		12	4	.028	3	.009
		11	3	.024	2	.007
		10	2	.018	1	.004
		9	2	.040	0	.002
		8	1	.024	0	.004
		7	0	.010	0	.010
		6	0	.022	-	
		5	0	.047	-	
	9	14	6	.047	4	.004
		13	4	.018	3	.005
		12	3	.017	2	.004
		11	3	.042	1	.002
		10	2	.029	1	.007
		9	1	.017	0	.002
		8	1	.036	0	.006
		7	0	.014	-	
		6	0	.030	-	

Table 46 CRITICAL VALUES FOR FISHER'S EXACT TEST FOR TWO BY TWO TABLES
(Description: pp. 130-131)

A	B	a	α ≤ 0.05 b	α ≤ 0.05 P	α ≤ 0.01 b	α ≤ 0.01 P
14	8	14	5	.036	4	.010
		13	4	.039	2	.002
		12	3	.032	2	.008
		11	2	.022	1	.005
		10	2	.048	0	.002
		9	1	.026	0	.004
		8	0	.009	0	.009
		7	0	.020	-	
		6	0	.040	-	
	7	14	4	.026	3	.006
		13	3	.025	2	.006
		12	2	.017	1	.003
		11	2	.041	1	.009
		10	1	.021	0	.003
		9	1	.043	0	.007
		8	0	.015	-	
		7	0	.030	-	
	6	14	3	.018	2	.003
		13	2	.014	1	.002
		12	2	.037	1	.007
		11	1	.018	0	.002
		10	1	.038	0	.005
		9	0	.012	-	
		8	0	.024	-	
		7	0	.044	-	
	5	14	2	.010	1	.001
		13	2	.037	1	.006
		12	1	.017	0	.002
		11	1	.038	0	.005
		10	0	.011	-	
		9	0	.022	-	
		8	0	.040	-	
	4	14	2	.039	1	.005
		13	1	.019	0	.002
		12	1	.044	0	.005
		11	0	.011	-	
		10	0	.023	-	
		9	0	.041	-	
	3	14	1	.022	0	.001
		13	0	.006	0	.006
		12	0	.015	-	
		11	0	.029	-	
	2	14	0	.008	0	.008
		13	0	.025	-	

A	B	a	α ≤ 0.05 b	α ≤ 0.05 P	α ≤ 0.01 b	α ≤ 0.01 P
15	15	15	11	.050	9	.008
		14	9	.040	7	.007
		13	7	.025	5	.004
		12	6	.030	4	.005
		11	5	.033	3	.005
		10	4	.033	2	.004
		9	3	.030	1	.003
		8	2	.025	1	.007
		7	1	.018	0	.003
		6	1	.040	0	.008
		5	0	.021	-	
		4	0	.050	-	
	14	15	10	.042	8	.006
		14	8	.031	6	.005
		13	7	.041	5	.007
		12	6	.046	4	.007
		11	5	.048	3	.007
		10	4	.046	2	.006
		9	3	.041	1	.004
		8	2	.033	1	.009
		7	1	.022	0	.004
		6	1	.049	-	
		5	0	.025	-	
	13	15	9	.035	7	.005
		14	7	.023	6	.009
		13	6	.029	4	.004
		12	5	.031	3	.004
		11	4	.030	2	.003
		10	3	.026	2	.008
		9	2	.020	1	.005
		8	2	.043	0	.002
		7	1	.029	0	.005
		6	0	.013	-	
		5	0	.031	-	
	12	15	8	.028	7	.010
		14	7	.043	5	.006
		13	6	.049	4	.007
		12	5	.049	3	.006
		11	4	.045	2	.005
		10	3	.038	1	.003
		9	2	.028	1	.007
		8	1	.018	0	.003
		7	1	.038	0	.007
		6	0	.017	-	
		5	0	.037	-	

Table 46 CRITICAL VALUES FOR FISHER'S EXACT TEST FOR TWO BY TWO TABLES
(Description: pp. 130-131)

			$\alpha \le 0.05$		$\alpha \le 0.01$	
A	B	a	b	P	b	P
15	11	15	7	.022	6	.007
		14	6	.032	4	.003
		13	5	.034	3	.003
		12	4	.032	2	.003
		11	3	.026	2	.008
		10	2	.019	1	.004
		9	2	.040	0	.002
		8	1	.024	0	.004
		7	1	.049	0	.010
		6	0	.022	-	
		5	0	.046	-	
	10	15	6	.017	5	.005
		14	5	.023	4	.007
		13	4	.022	3	.007
		12	3	.018	2	.005
		11	3	.042	1	.003
		10	2	.029	1	.007
		9	1	.016	0	.002
		8	1	.034	0	.006
		7	0	.013	-	
		6	0	.028	-	
	9	15	6	.042	4	.003
		14	5	.047	3	.004
		13	4	.042	2	.003
		12	3	.032	2	.009
		11	2	.021	1	.005
		10	2	.045	0	.002
		9	1	.024	0	.004
		8	1	.048	0	.009
		7	0	.019	-	
		6	0	.037	-	
	8	15	5	.032	4	.008
		14	4	.033	3	.009
		13	3	.026	2	.006
		12	2	.017	1	.003
		11	2	.037	1	.008
		10	1	.019	0	.003
		9	1	.038	0	.006
		8	0	.013	-	
		7	0	.026	-	
		6	0	.050	-	

			$\alpha \le 0.05$		$\alpha \le 0.01$	
A	B	a	b	P	b	P
15	7	15	4	.023	3	.005
		14	3	.021	2	.004
		13	2	.014	1	.002
		12	2	.032	1	.007
		11	1	.015	0	.002
		10	1	.032	0	.005
		9	0	.010	-	
		8	0	.020	-	
		7	0	.038	-	
	6	15	3	.015	2	.003
		14	2	.011	1	.002
		13	2	.031	1	.006
		12	1	.014	0	.002
		11	1	.029	0	.004
		10	0	.009	0	.009
		9	0	.017	-	
		8	0	.032	-	
	5	15	2	.009	2	.009
		14	2	.032	1	.005
		13	1	.014	0	.001
		12	1	.031	0	.004
		11	0	.008	0	.008
		10	0	.016	-	
		9	0	.030	-	
	4	15	2	.035	1	.004
		14	1	.016	0	.001
		13	1	.037	0	.004
		12	0	.009	0	.009
		11	0	.018	-	
		10	0	.033	-	
	3	15	1	.020	0	.001
		14	0	.005	0	.005
		13	0	.012	-	
		12	0	.025	-	
		11	0	.043	-	
	2	15	0	.007	0	.007
		14	0	.022	-	
		13	0	.044	-	

Table 46 CRITICAL VALUES FOR FISHER'S EXACT TEST FOR TWO BY TWO TABLES
(Description: pp. 130-131)

A	B	a	α ≤ 0.05 b	α ≤ 0.05 P	α ≤ 0.01 b	α ≤ 0.01 P
16	16	16	11	.022	10	.009
		15	10	.041	8	.008
		14	8	.027	6	.005
		13	7	.033	5	.006
		12	6	.037	4	.006
		11	5	.038	3	.006
		10	4	.037	2	.005
		9	3	.033	1	.003
		8	2	.027	1	.008
		7	1	.019	0	.003
		6	1	.041	0	.009
		5	0	.022	–	
	15	16	11	.043	9	.007
		15	9	.033	7	.005
		14	8	.044	6	.008
		13	6	.023	5	.009
		12	5	.024	4	.009
		11	4	.023	3	.008
		10	4	.049	2	.006
		9	3	.043	1	.004
		8	2	.035	0	.002
		7	1	.023	0	.004
		6	0	.011	–	
		5	0	.026	–	
	14	16	10	.037	8	.005
		15	8	.025	7	.010
		14	7	.032	5	.005
		13	6	.035	4	.005
		12	5	.035	3	.005
		11	4	.033	2	.004
		10	3	.028	2	.009
		9	2	.021	1	.006
		8	2	.045	0	.002
		7	1	.030	0	.006
		6	0	.013	–	
		5	0	.031	–	
	13	16	9	.030	7	.004
		15	8	.047	6	.007
		14	6	.023	5	.008
		13	5	.023	4	.008
		12	4	.022	3	.007
		11	4	.048	2	.005
		10	3	.039	1	.003
		9	2	.029	1	.008
		8	1	.018	0	.003
		7	1	.038	0	.007
		6	0	.017	–	
		5	0	.037	–	
16	12	16	8	.024	7	.008
		15	7	.036	5	.004
		14	6	.040	4	.005
		13	5	.039	3	.004
		12	4	.034	2	.003
		11	3	.027	2	.008
		10	2	.019	1	.005
		9	2	.040	0	.002
		8	1	.024	0	.004
		7	1	.048	0	.010
		6	0	.021	–	
		5	0	.044	–	
	11	16	7	.019	6	.006
		15	6	.027	5	.009
		14	5	.027	4	.009
		13	4	.024	3	.008
		12	3	.019	2	.005
		11	3	.041	1	.003
		10	2	.028	1	.007
		9	1	.016	0	.002
		8	1	.033	0	.006
		7	0	.013	–	
		6	0	.027	–	
	10	16	7	.046	5	.004
		15	5	.018	4	.005
		14	4	.017	3	.005
		13	4	.042	2	.003
		12	3	.032	2	.009
		11	2	.021	1	.005
		10	2	.042	0	.002
		9	1	.023	0	.004
		8	1	.045	0	.008
		7	0	.017	–	
		6	0	.035	–	
	9	16	6	.037	5	.010
		15	5	.040	3	.003
		14	4	.034	3	.010
		13	3	.025	2	.007
		12	2	.016	1	.003
		11	2	.033	1	.008
		10	1	.017	0	.002
		9	1	.034	0	.006
		8	0	.012	–	
		7	0	.024	–	
		6	0	.045	–	
	8	16	5	.028	4	.007
		15	4	.028	3	.007
		14	3	.021	2	.005
		13	3	.047	1	.002
		12	2	.028	1	.006

Table 46 CRITICAL VALUES FOR FISHER'S EXACT TEST FOR TWO BY TWO TABLES
(Description: pp. 130-131)

			$\alpha \leq 0.05$		$\alpha \leq 0.01$	
A	B	a	b	P	b	P
16	8	11	1	.014	0	.002
		10	1	.027	0	.004
		9	0	.009	0	.009
		8	0	.017	-	
		7	0	.033	-	
17	17	17	12	.022	11	.009
		16	11	.043	9	.008
		15	9	.029	7	.005
		14	8	.035	6	.007
		13	7	.040	5	.007
		12	6	.042	4	.007
		11	5	.042	3	.007
		10	4	.040	2	.005
		9	3	.035	1	.003
		8	2	.029	1	.008
		7	1	.020	0	.004
		6	1	.043	0	.009
		5	0	.022	-	
	16	17	12	.044	10	.007
		16	10	.035	8	.006
		15	9	.046	7	.009
		14	7	.025	5	.004
		13	6	.027	4	.004
		12	5	.027	3	.004
		11	4	.025	3	.009
		10	3	.022	2	.007
		9	3	.046	1	.004
		8	2	.036	0	.002
		7	1	.024	0	.005
		6	0	.011	-	
		5	0	.026	-	
	15	17	11	.038	9	.006
		16	9	.027	7	.004
		15	8	.035	6	.006
		14	7	.040	5	.006
		13	6	.041	4	.006
		12	5	.039	3	.005
		11	4	.035	2	.004
		10	3	.029	2	.010
		9	2	.022	1	.006
		8	2	.046	0	.002
		7	1	.030	0	.006
		6	0	.014	-	
		5	0	.031	-	
	14	17	10	.032	8	.004
		16	8	.021	7	.008
		15	7	.026	6	.010
		14	6	.028	4	.004
17	14	13	5	.027	4	.010
		12	4	.024	3	.008
		11	4	.049	2	.006
		10	3	.040	1	.003
		9	2	.029	1	.008
		8	1	.018	0	.003
		7	1	.038	0	.007
		6	0	.017	-	
		5	0	.036	-	
	13	17	9	.026	8	.009
		16	8	.040	6	.005
		15	7	.045	5	.006
		14	6	.045	4	.006
		13	5	.042	3	.005
		12	4	.035	2	.004
		11	3	.028	2	.009
		10	2	.019	1	.005
		9	2	.040	0	.002
		8	1	.024	0	.004
		7	1	.047	0	.010
		6	0	.021	-	
		5	0	.043	-	
	12	17	8	.021	7	.007
		16	7	.030	5	.003
		15	6	.033	4	.004
		14	5	.030	3	.003
		13	4	.026	3	.008
		12	3	.020	2	.006
		11	3	.041	1	.003
		10	2	.028	1	.007
		9	1	.016	0	.002
		8	1	.032	0	.006
		7	0	.012	-	
		6	0	.026	-	
	11	17	7	.016	6	.005
		16	6	.022	5	.007
		15	5	.022	4	.007
		14	4	.019	3	.006
		13	4	.042	2	.004
		12	3	.031	2	.009
		11	2	.020	1	.005
		10	2	.040	0	.001
		9	1	.022	0	.004
		8	1	.042	0	.008
		7	0	.016	-	
		6	0	.033	-	
	10	17	7	.041	5	.003
		16	6	.047	4	.004
		15	5	.043	3	.004
		14	4	.034	2	.002

Table 46 CRITICAL VALUES FOR FISHER'S EXACT TEST FOR TWO BY TWO TABLES
(Description: pp. 130-131)

A	B	a	$\alpha \le 0.05$		$\alpha \le 0.01$	
			b	P	b	P
17	10	13	3	.024	2	.007
		12	3	.049	1	.003
		11	2	.031	1	.007
		10	1	.016	0	.002
		9	1	.031	0	.005
		8	0	.011	–	
		7	0	.022	–	
		6	0	.042	–	
	9	17	6	.032	5	.008
		16	5	.034	4	.010
		15	4	.028	3	.008
		14	3	.020	2	.005
		13	3	.042	1	.002
		12	2	.025	1	.006
		11	2	.048	0	.002
		10	1	.024	0	.004
		9	1	.045	0	.008
		8	0	.016	–	
		7	0	.030	–	
	8	17	5	.024	4	.006
		16	4	.023	3	.006
		15	3	.017	2	.004
		14	3	.039	2	.010
		13	2	.022	1	.004
		12	2	.043	1	.010
		11	1	.020	0	.003
		10	1	.038	0	.006
		9	0	.012	–	
		8	0	.022	–	
		7	0	.040	–	
18	18	18	13	.023	12	.010
		17	12	.044	10	.009
		16	10	.030	8	.006
		15	9	.038	7	.008
		14	8	.043	6	.009
		13	7	.046	5	.009
		12	6	.047	4	.009
		11	5	.046	3	.008
		10	4	.043	2	.006
		9	3	.038	1	.004
		8	2	.030	1	.009
		7	1	.020	0	.004
		6	1	.044	0	.010
		5	0	.023	–	
	17	18	13	.045	11	.008
		17	11	.036	9	.007
		16	10	.049	8	.010
		15	8	.028	6	.005
18	17	14	7	.030	5	.005
		13	6	.031	4	.005
		12	5	.030	3	.004
		11	4	.028	2	.003
		10	3	.023	2	.008
		9	3	.047	1	.005
		8	2	.037	0	.002
		7	1	.025	0	.005
		6	0	.011	–	
		5	0	.026	–	
	16	18	12	.039	10	.006
		17	10	.029	8	.005
		16	9	.038	7	.007
		15	8	.043	6	.008
		14	7	.046	5	.008
		13	6	.045	4	.007
		12	5	.042	3	.006
		11	4	.037	2	.004
		10	3	.031	1	.003
		9	2	.023	1	.006
		8	2	.046	0	.002
		7	1	.030	0	.006
		6	0	.014	–	
		5	0	.031	–	
	15	18	11	.033	9	.005
		17	9	.023	8	.009
		16	8	.029	6	.004
		15	7	.031	5	.005
		14	6	.031	4	.004
		13	5	.029	3	.004
		12	4	.025	3	.009
		11	3	.020	2	.006
		10	3	.041	1	.004
		9	2	.030	1	.008
		8	1	.018	0	.003
		7	1	.038	0	.007
		6	0	.017	–	
		5	0	.036	–	
	14	18	10	.028	9	.010
		17	9	.043	7	.006
		16	8	.050	6	.008
		15	6	.022	5	.008
		14	6	.049	4	.007
		13	5	.044	3	.006
		12	4	.037	2	.004
		11	3	.028	2	.009
		10	2	.020	1	.005
		9	2	.039	0	.002
		8	1	.024	0	.004

Table 46 CRITICAL VALUES FOR FISHER'S EXACT TEST FOR TWO BY TWO TABLES
(Description: pp. 130-131)

A	B	a	α ≤ 0.05 b	α ≤ 0.05 P	α ≤ 0.01 b	α ≤ 0.01 P
18	14	7	1	.047	0	.009
		6	0	.020	-	
		5	0	.043	-	
	13	18	9	.023	8	.008
		17	8	.034	6	.004
		16	7	.037	5	.005
		15	6	.036	4	.004
		14	5	.032	3	.004
		13	4	.027	3	.009
		12	3	.020	2	.006
		11	3	.040	1	.003
		10	2	.027	1	.007
		9	1	.015	0	.002
		8	1	.031	0	.006
		7	0	.012	-	
		6	0	.025	-	
	12	18	8	.018	7	.006
		17	7	.026	6	.009
		16	6	.027	5	.009
		15	5	.024	4	.008
		14	4	.020	3	.006
		13	4	.042	2	.004
		12	3	.030	2	.009
		11	2	.019	1	.005
		10	2	.038	0	.001
		9	1	.021	0	.003
		8	1	.040	0	.007
		7	0	.016	-	
		6	0	.031	-	
	11	18	8	.045	6	.004
		17	6	.018	5	.006
		16	5	.018	4	.005
		15	5	.043	3	.004
		14	4	.033	2	.003
		13	3	.023	2	.007
		12	3	.046	1	.003
		11	2	.029	1	.007
		10	1	.015	0	.002
		9	1	.029	0	.005
		8	0	.010	-	
		7	0	.020	-	
		6	0	.039	-	
	10	18	7	.037	5	.003
		17	6	.041	4	.003
		16	5	.036	3	.003
		15	4	.028	3	.008
		14	3	.019	2	.005
		13	3	.039	1	.002
		12	2	.023	1	.005
		11	2	.043	0	.001
18	10	10	1	.022	0	.003
		9	1	.040	0	.007
		8	0	.014	-	
		7	0	.027	-	
		6	0	.049	-	
	9	18	6	.029	5	.007
		17	5	.030	4	.008
		16	4	.023	3	.006
		15	3	.016	2	.004
		14	3	.034	2	.009
		13	2	.019	1	.004
		12	2	.037	1	.009
		11	1	.018	0	.002
		10	1	.033	0	.005
		9	0	.010	-	
		8	0	.020	-	
		7	0	.036	-	
19	19	19	14	.023	13	.010
		18	13	.045	11	.009
		17	11	.031	9	.006
		16	10	.039	8	.009
		15	9	.046	6	.004
		14	8	.050	5	.004
		13	6	.025	4	.004
		12	5	.024	3	.003
		11	5	.050	3	.009
		10	4	.046	2	.006
		9	3	.039	1	.004
		8	2	.031	1	.009
		7	1	.021	0	.004
		6	1	.045	0	.010
		5	0	.023	-	
	18	19	14	.046	12	.008
		18	12	.037	10	.007
		17	10	.024	8	.004
		16	9	.030	7	.006
		15	8	.033	6	.006
		14	7	.035	5	.006
		13	6	.035	4	.006
		12	5	.033	3	.005
		11	4	.030	2	.004
		10	3	.025	2	.008
		9	3	.049	1	.005
		8	2	.038	0	.002
		7	1	.025	0	.005
		6	0	.012	-	
		5	0	.027	-	

Table 46 CRITICAL VALUES FOR FISHER'S EXACT TEST FOR TWO BY TWO TABLES (Description: pp. 130-131)

A	B	a	α ≤ 0.05		α ≤ 0.01	
			b	P	b	P
19	17	19	13	.040	11	.006
		18	11	.030	9	.005
		17	10	.040	8	.008
		16	9	.047	7	.009
		15	8	.050	6	.010
		14	6	.023	5	.010
		13	6	.049	4	.008
		12	5	.045	3	.007
		11	4	.039	2	.005
		10	3	.032	1	.003
		9	2	.024	1	.007
		8	2	.047	0	.002
		7	1	.031	0	.006
		6	0	.014	–	
		5	0	.031	–	
	16	19	12	.035	10	.005
		18	10	.024	9	.010
		17	9	.031	7	.005
		16	8	.035	6	.006
		15	7	.036	5	.006
		14	6	.034	4	.005
		13	5	.031	3	.004
		12	4	.027	3	.010
		11	3	.021	2	.007
		10	3	.042	1	.004
		9	2	.030	1	.009
		8	1	.018	0	.003
		7	1	.037	0	.007
		6	0	.017	–	
		5	0	.036	–	
	15	19	11	.029	9	.004
		18	10	.046	8	.007
		17	8	.023	7	.009
		16	7	.025	6	.010
		15	6	.024	5	.009
		14	5	.022	4	.008
		13	5	.045	3	.006
		12	4	.037	2	.004
		11	3	.029	2	.009
		10	2	.020	1	.005
		9	2	.039	0	.002
		8	1	.023	0	.004
		7	1	.046	0	.009
		6	0	.020	–	
		5	0	.042	–	
	14	19	10	.024	9	.008
		18	9	.037	7	.005
		17	8	.042	6	.006
		16	7	.042	5	.006
		15	6	.039	4	.005
		14	5	.034	3	.004

A	B	a	α ≤ 0.05		α ≤ 0.01	
			b	P	b	P
19	14	13	4	.027	3	.009
		12	3	.020	2	.006
		11	3	.040	1	.003
		10	2	.027	1	.007
		9	1	.016	0	.002
		8	1	.030	0	.005
		7	0	.012	–	
		6	0	.024	–	
		5	0	.049	–	
	13	19	9	.020	8	.006
		18	8	.029	6	.003
		17	7	.031	5	.004
		16	6	.029	4	.003
		15	5	.025	4	.009
		14	4	.020	3	.006
		13	4	.041	2	.004
		12	3	.029	2	.009
		11	2	.019	1	.005
		10	2	.036	1	.010
		9	1	.020	0	.003
		8	1	.038	0	.007
		7	0	.015	–	
		6	0	.030	–	
	12	19	9	.049	7	.005
		18	7	.022	6	.007
		17	6	.022	5	.007
		16	5	.019	4	.006
		15	5	.042	3	.004
		14	4	.032	2	.003
		13	3	.023	2	.006
		12	3	.043	1	.003
		11	2	.027	1	.007
		10	2	.050	0	.002
		9	1	.027	0	.005
		8	1	.050	0	.010
		7	0	.019	–	
		6	0	.037	–	
	11	19	8	.041	6	.003
		18	7	.047	5	.004
		17	6	.043	4	.004
		16	5	.035	3	.003
		15	4	.027	3	.008
		14	3	.018	2	.005
		13	3	.035	1	.002
		12	2	.021	1	.005
		11	2	.040	0	.001
		10	1	.020	0	.003
		9	1	.037	0	.006
		8	0	.013	–	
		7	0	.025	–	
		6	0	.046	–	

Table 46 CRITICAL VALUES FOR FISHER'S EXACT TEST FOR TWO BY TWO TABLES (Description: pp. 130-131)

			α ≤ 0.05		α ≤ 0.01	
A	B	a	b	P	b	P
19	10	19	7	.033	6	.009
		18	6	.036	4	.003
		17	5	.030	4	.009
		16	4	.022	3	.006
		15	4	.047	2	.004
		14	3	.030	2	.008
		13	2	.017	1	.004
		12	2	.033	1	.008
		11	1	.016	0	.002
		10	1	.029	0	.005
		9	0	.009	0	.009
		8	0	.018	-	
		7	0	.032	-	
	9	19	6	.026	5	.006
		18	5	.026	4	.007
		17	4	.020	3	.005
		16	4	.044	2	.003
		15	3	.028	2	.007
		14	2	.015	1	.003
		13	2	.029	1	.006
		12	1	.013	0	.002
		11	1	.024	0	.004
		10	1	.042	0	.007
		9	0	.013	-	
		8	0	.024	-	
		7	0	.043	-	
20	20	20	15	.024	13	.004
		19	14	.046	12	.010
		18	12	.032	10	.007
		17	11	.041	9	.009
		16	10	.048	7	.005
		15	8	.027	6	.005
		14	7	.028	5	.005
		13	6	.028	4	.005
		12	5	.027	3	.004
		11	4	.024	3	.009
		10	4	.048	2	.007
		9	3	.041	1	.004
		8	2	.032	1	.010
		7	1	.022	0	.004
		6	1	.046	-	
		5	0	.024	-	
	19	20	15	.047	13	.008
		19	13	.039	11	.008
		18	11	.026	9	.005
		17	10	.032	8	.006
		16	9	.036	7	.007
		15	8	.038	6	.008
		14	7	.039	5	.007
		13	6	.038	4	.007

			α ≤ 0.05		α ≤ 0.01	
A	B	a	b	P	b	P
20	19	12	5	.035	3	.005
		11	4	.031	2	.004
		10	3	.026	2	.009
		9	2	.019	1	.005
		8	2	.039	0	.002
		7	1	.026	0	.005
		6	0	.012	-	
		5	0	.027	-	
	18	20	14	.041	12	.007
		19	12	.032	10	.006
		18	11	.043	9	.008
		17	10	.050	7	.004
		16	8	.026	6	.005
		15	7	.027	5	.004
		14	6	.026	4	.004
		13	5	.024	4	.009
		12	5	.047	3	.007
		11	4	.041	2	.005
		10	3	.033	1	.003
		9	2	.024	1	.007
		8	2	.048	0	.003
		7	1	.031	0	.006
		6	0	.014	-	
		5	0	.031	-	
	17	20	13	.036	11	.005
		19	11	.026	9	.004
		18	10	.034	8	.006
		17	9	.038	7	.007
		16	8	.040	6	.007
		15	7	.039	5	.007
		14	6	.037	4	.006
		13	5	.033	3	.005
		12	4	.028	2	.003
		11	3	.022	2	.007
		10	3	.042	1	.004
		9	2	.031	1	.009
		8	1	.019	0	.003
		7	1	.037	0	.008
		6	0	.017	-	
		5	0	.036	-	
	16	20	12	.031	10	.004
		19	11	.049	9	.008
		18	9	.026	7	.004
		17	8	.028	6	.004
		16	7	.028	5	.004
		15	6	.026	4	.004
		14	5	.023	4	.009
		13	5	.046	3	.007
		12	4	.038	2	.004
		11	3	.029	2	.010

Table 46 CRITICAL VALUES FOR FISHER'S EXACT TEST FOR TWO BY TWO TABLES
(Description: pp. 130-131)

A	B	a	α ≤ 0.05 b	α ≤ 0.05 P	α ≤ 0.01 b	α ≤ 0.01 P
20	16	10	2	.020	1	.005
		9	2	.039	0	.002
		8	1	.023	0	.004
		7	1	.045	0	.009
		6	0	.020	–	
		5	0	.041	–	
	15	20	11	.026	10	.009
		19	10	.040	8	.006
		18	9	.046	7	.007
		17	8	.047	6	.008
		16	7	.045	5	.007
		15	6	.040	4	.006
		14	5	.034	3	.004
		13	4	.028	3	.010
		12	3	.020	2	.006
		11	3	.039	1	.003
		10	2	.026	1	.007
		9	2	.049	0	.002
		8	1	.029	0	.005
		7	0	.012	–	
		6	0	.024	–	
		5	0	.048	–	
	14	20	10	.022	9	.007
		19	9	.032	7	.004
		18	8	.035	6	.005
		17	7	.035	5	.005
		16	6	.031	4	.004
		15	5	.026	4	.009
		14	4	.020	3	.007
		13	4	.040	2	.004
		12	3	.029	2	.009
		11	2	.018	1	.005
		10	2	.035	1	.010
		9	1	.019	0	.003
		8	1	.037	0	.007
		7	0	.014	–	
		6	0	.029	–	
	13	20	9	.017	8	.005
		19	8	.025	7	.008
		18	7	.026	6	.009
		17	6	.024	5	.008
		16	5	.020	4	.007
		15	5	.041	3	.005
		14	4	.031	2	.003
		13	3	.022	2	.006
		12	3	.041	1	.003
		11	2	.026	1	.007
		10	2	.047	0	.002
20	13	9	1	.026	0	.004
		8	1	.047	0	.009
		7	0	.018	–	
		6	0	.035	–	
	12	20	9	.044	7	.004
		19	7	.018	6	.006
		18	6	.018	5	.006
		17	6	.043	4	.005
		16	5	.034	3	.003
		15	4	.025	3	.008
		14	4	.049	2	.005
		13	3	.033	2	.010
		12	2	.020	1	.005
		11	2	.036	1	.009
		10	1	.018	0	.003
		9	1	.034	0	.006
		8	0	.012	–	
		7	0	.023	–	
		6	0	.043	–	
	11	20	8	.037	6	.003
		19	7	.042	5	.004
		18	6	.037	4	.003
		17	5	.029	4	.009
		16	4	.021	3	.006
		15	4	.042	2	.003
		14	3	.028	2	.008
		13	2	.016	1	.003
		12	2	.029	1	.007
		11	1	.014	0	.002
		10	1	.026	0	.004
		9	1	.046	0	.008
		8	0	.016	–	
		7	0	.029	–	
	10	20	7	.030	6	.008
		19	6	.031	5	.009
		18	5	.026	4	.007
		17	4	.018	3	.005
		16	4	.039	2	.003
		15	3	.024	2	.006
		14	3	.045	1	.003
		13	2	.025	1	.006
		12	2	.045	0	.001
		11	1	.021	0	.003
		10	1	.037	0	.006
		9	0	.012	–	
		8	0	.022	–	
		7	0	.038	–	

Table 47 SAMPLE SIZE TO COMPARE TWO PROPORTIONS
(Table: p. 147)

A one-sided Fisher's exact test is to be used to test equality of two proportions with equal numbers in each group. If the true proportions are P_1 and P_2, this table gives approximately the sample size needed for each group when the significance level is α and probability of type II error is β. The arcsine approximation is used:

$$n = \frac{1}{2}\left(\frac{z_\alpha + z_\beta}{\arcsin\sqrt{P_1} - \arcsin\sqrt{P_2}}\right)^2$$

where z_γ is given by

$$\int_{z_\gamma}^{\infty} \frac{\exp(-x^2/2)}{\sqrt{2\pi}}\, dx = \gamma$$

The angles are in radians.

Table 47 SAMPLE SIZE TO COMPARE TWO PROPORTIONS
(Description: p. 146)

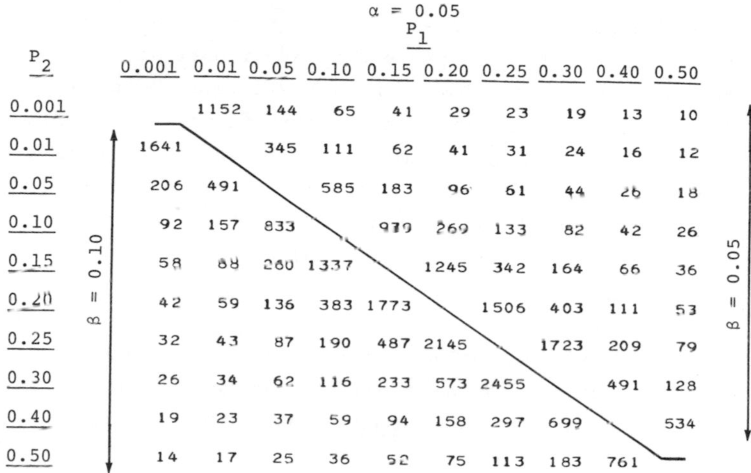

$\alpha = 0.05$

Below diagonal: $\beta = 0.10$; above diagonal: $\beta = 0.05$

P_2 \ P_1	0.001	0.01	0.05	0.10	0.15	0.20	0.25	0.30	0.40	0.50
0.001		1152	144	65	41	29	23	19	13	10
0.01	1641		345	111	62	41	31	24	16	12
0.05	206	491		585	183	96	61	44	26	18
0.10	92	157	833		979	269	133	82	42	26
0.15	58	88	260	1337		1245	342	164	66	36
0.20	42	59	136	383	1773		1506	403	111	53
0.25	32	43	87	190	487	2145		1723	209	79
0.30	26	34	62	116	233	573	2455		491	128
0.40	19	23	37	59	94	158	297	699		534
0.50	14	17	25	36	52	75	113	183	761	

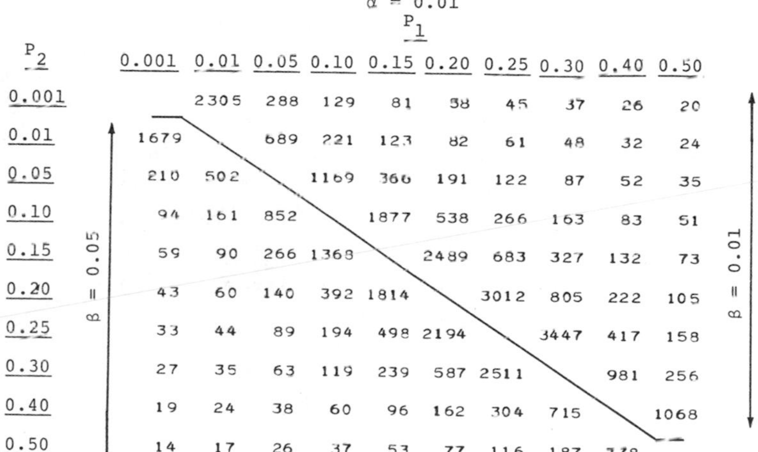

$\alpha = 0.01$

Below diagonal: $\beta = 0.05$; above diagonal: $\beta = 0.01$

P_2 \ P_1	0.001	0.01	0.05	0.10	0.15	0.20	0.25	0.30	0.40	0.50
0.001		2305	288	129	81	58	45	37	26	20
0.01	1679		689	221	123	82	61	48	32	24
0.05	210	502		1169	366	191	122	87	52	35
0.10	94	161	852		1877	538	266	163	83	51
0.15	59	90	266	1368		2489	683	327	132	73
0.20	43	60	140	392	1814		3012	805	222	105
0.25	33	44	89	194	498	2194		3447	417	158
0.30	27	35	63	119	239	587	2511		981	256
0.40	19	24	38	60	96	162	304	715		1068
0.50	14	17	26	37	53	77	116	187	778	

Table 48 CRITICAL VALUES FOR A NONPARAMETRIC TEST OF LOCATION (Table: pp. 149-150)

Let two independent samples of sizes m and n be drawn from continuous distributions that are identical, except possibly for a location parameter. The hypothesis that the distributions are identical may be tested by counting the number of points in the sample of size m that are larger than all points in the sample of size n. The hypothesis is rejected if this number is larger than or equal to the critical value determined from the table.

To find the critical value for a given m and n, go across the row labeled n to the first number greater than or equal to m. The column heading of this number is the critical value.

For large and approximately equal sample sizes the critical values are 5 for the 5% test and 7 for the 1% test.

Example: Suppose that m = 40 and n = 22, and a test at $\alpha = 0.05$ is to be performed. On the $\alpha = 0.05$ portion of the table, go across the row labeled 22. The first number greater than or equal to 40 is 44, and the column heading is 7. Thus, the critical value is 7. Reject equality of the distributions if 7 or more points of the sample of size 40 are larger than all points in the sample of size 22.

Table 48 CRITICAL VALUES FOR A NONPARAMETRIC TEST OF LOCATION (Description: p. 148)

α = 0.01 critical values

n	1	2	3	4	5	6	7	8	9	10	11	12	13	14	15
6	-	-	-	4	6	8	9	11	13	15	17	19	20	22	24
7	-	-	3	4	6	8	10	12	14	17	19	21	23	25	27
8	-	-	3	5	7	9	11	14	16	18	20	23	25	27	29
9	-	-	3	5	8	10	12	15	17	20	22	25	27	30	32
10	-	-	3	6	8	11	14	16	19	21	24	27	30	32	35
11	-	-	4	6	9	12	15	17	20	23	26	29	32	35	38
12	-	-	4	7	10	13	16	19	22	25	28	31	34	37	40
13	-	2	4	7	10	13	17	20	23	27	30	33	36	40	43
14	-	2	4	8	11	14	18	21	25	28	32	35	39	42	46
15	-	2	5	8	12	15	19	22	26	30	34	37	41	45	49
16	-	2	5	9	12	16	20	24	28	32	36	39	43	47	50
17	-	2	5	9	13	17	21	25	29	33	37	42	46	50	
18	-	2	6	9	14	18	22	26	31	35	39	44	48	50	
19	-	2	6	10	14	19	23	28	32	37	41	46	50		
20	-	2	6	10	15	19	24	29	34	38	43	48	50		
21	-	2	6	11	15	20	25	30	35	40	45	50			
22	-	2	7	11	16	21	26	31	37	42	47	50			
23	-	3	7	12	17	22	27	33	38	43	49	50			
24	-	3	7	12	17	23	28	34	40	45	50				
25	-	3	7	13	18	24	29	35	41	47	50				
26	-	3	8	13	19	25	31	37	43	49	50				
27	-	3	8	14	19	25	32	38	44	50					
28	-	3	8	14	20	26	33	39	46	50					
29	-	3	9	14	21	27	34	40	47	50					
30	-	3	9	15	21	28	35	42	49	50					
31	-	3	9	15	22	29	36	43	50						
32	-	4	9	16	23	30	37	44	50						
33	-	4	10	16	23	31	38	45	50						
34	-	4	10	17	24	32	39	47	50						
35	-	4	10	17	25	32	40	48	50						
36	-	4	10	18	25	33	41	49	50						
37	-	4	11	18	26	34	42	50							
38	-	4	11	19	27	35	43	50							
39	-	4	11	19	27	36	44	50							
40	-	4	12	20	28	37	46	50							
41	-	5	12	20	29	38	47	50							
42	-	5	12	20	29	38	48	50							
43	-	5	12	21	30	39	49	50							
44	-	5	13	21	31	40	50								
45	-	5	13	22	31	41	50								
46	-	5	13	22	32	42	50								
47	-	5	13	23	33	43	50								
48	-	5	14	23	33	44	50								
49	-	5	14	24	34	44	50								
50	-	6	14	24	35	45	50								

Table 48 CRITICAL VALUES FOR A NONPARAMETRIC TEST OF LOCATION (Description: p. 148)

$\alpha = 0.05$ critical values

n	1	2	3	4	5	6	7	8	9	10	11	12	13	14	15
6	-	2	4	7	9	12	14	17	19	22	24	27	29	32	34
7	-	2	5	7	10	13	16	19	22	24	27	30	33	36	39
8	-	2	5	8	11	15	18	21	24	27	30	34	37	40	43
9	-	3	6	9	13	16	20	23	27	30	34	37	41	44	48
10	-	3	6	10	14	18	21	25	29	33	37	41	44	48	50
11	-	3	7	11	15	19	23	27	32	36	40	44	48	50	
12	-	3	8	12	16	21	25	30	34	39	43	48	50		
13	-	4	8	13	17	22	27	32	37	41	46	50			
14	-	4	9	14	19	24	29	34	39	44	49	50			
15	-	4	9	15	20	25	31	36	42	47	50				
16	-	5	10	15	21	27	33	38	44	50					
17	-	5	10	16	22	28	34	41	47	50					
18	-	5	11	17	24	30	36	43	49	50					
19	1	5	12	18	25	31	38	45	50						
20	1	6	12	19	26	33	40	47	50						
21	1	6	13	20	27	35	42	49	50						
22	1	6	13	21	28	36	44	50							
23	1	7	14	22	30	38	46	50							
24	1	7	15	23	31	39	48	50							
25	1	7	15	23	32	41	49	50							
26	1	8	16	24	33	42	50								
27	1	8	16	25	34	44	50								
28	1	8	17	26	36	45	50								
29	1	8	17	27	37	47	50								
30	1	9	18	28	38	48	50								
31	1	9	19	29	39	50									
32	1	9	19	30	41	50									
33	1	10	20	31	42	50									
34	1	10	20	32	43	50									
35	1	10	21	32	44	50									
36	1	10	22	33	45	50									
37	1	11	22	34	47	50									
38	1	11	23	35	48	50									
39	2	11	23	36	49	50									
40	2	12	24	37	50										
41	2	12	24	38	50										
42	2	12	25	39	50										
43	2	12	26	40	50										
44	2	13	26	40	50										
45	2	13	27	41	50										
46	2	13	27	42	50										
47	2	14	28	43	50										
48	2	14	29	44	50										
49	2	14	29	45	50										
50	2	14	30	46	50										

Table 49 CRITICAL VALUES FOR A NONPARAMETRIC TEST OF DISPERSION (Table: pp. 152-153)

Given two independent samples of sizes $n \geq 2$ and $m \geq 2$ with equal medians that may or may not differ by a scale parameter, it is desired to test the null hypothesis of equal scale parameters against the alternative that the sample of size m has more dispersion. Count the number of points in the sample of size m which lie outside all points of the sample of size n. If this number of points is greater than or equal to the tabulated critical value, reject the null hypothesis. Critical values for $\alpha = 0.05$ and 0.01 are given.

To find the critical value for a given m and n, go across the row labeled n to the first number greater than or equal to m. The <u>column</u> heading of this number is the critical value. A dash indicates that m must be greater than or equal to the first value in the row for a critical value to exist.

An alternative test suggested by Siegel and Tukey is to order the combined samples and "rank" from the outside in, i.e.,

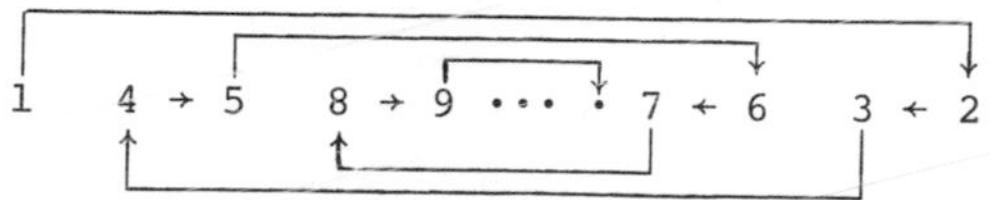

and to perform a one- or two-sided Mann-Whitney or Wilcoxon test (Table 28).

Table 49 CRITICAL VALUES FOR A NONPARAMETRIC TEST OF DISPERSION (Description: p. 151)

α = 0.01 critical values

n	2	3	4	5	6	7	8	9	10	11	12	13	14	15
6	-	-	-	-	6	7	8	10	11	13	14	15	17	18
7	-	-	-	5	6	8	9	11	12	14	15	17	18	20
8	-	-	-	5	7	8	10	12	13	15	17	18	20	22
9	-	-	4	5	7	9	11	12	14	16	18	20	21	23
10	-	-	4	6	8	10	11	13	15	17	19	21	23	25
11	-	-	4	6	8	10	12	14	16	18	21	23	25	27
12	-	3	5	7	9	11	13	15	17	20	22	24	26	29
13	-	3	5	7	9	12	14	16	19	21	23	26	28	30
14	-	3	5	7	10	12	15	17	20	22	25	27	30	32
15	-	3	5	8	10	13	15	18	21	23	26	29	31	34
16	-	3	6	8	11	14	16	19	22	25	27	30	33	36
17	-	3	6	9	11	14	17	20	23	26	29	32	35	38
18	-	4	6	9	12	15	18	21	24	27	30	33	36	40
19	-	4	6	9	12	16	19	22	25	28	32	35	38	41
20	-	4	7	10	13	16	20	23	26	30	33	36	40	43
21	-	4	7	10	14	17	20	24	27	31	34	38	42	45
22	-	4	7	11	14	18	21	25	29	32	36	40	43	47
23	2	4	8	11	15	18	22	26	30	33	37	41	45	49
24	2	4	8	11	15	19	23	27	31	35	39	43	47	50
25	2	5	8	12	16	20	24	28	32	36	40	44	48	50
26	2	5	8	12	16	20	25	29	33	37	42	46	50	
27	2	5	9	13	17	21	25	30	34	39	43	47	50	
28	2	5	9	13	17	22	26	31	35	40	44	49	50	
29	2	5	9	13	18	22	27	32	36	41	46	50		
30	2	5	10	14	18	23	28	33	37	42	47	50		
31	2	6	10	14	19	24	29	34	39	44	49	50		
32	2	6	10	15	20	25	30	35	40	45	50			
33	2	6	10	15	20	25	30	36	41	46	50			
34	2	6	11	16	21	26	31	37	42	47	50			
35	2	6	11	16	21	27	32	38	43	49	50			
36	2	6	11	16	22	27	33	39	44	50				
37	2	7	11	17	22	28	34	40	45	50				
38	2	7	12	17	23	29	35	40	46	50				
39	2	7	12	18	23	29	35	41	48	50				
40	3	7	12	18	24	30	36	42	49	50				
41	3	7	13	18	25	31	37	43	50					
42	3	7	13	19	25	31	38	44	50					
43	3	8	13	19	26	32	39	45	50					
44	3	8	13	20	26	33	40	46	50					
45	3	8	14	20	27	34	40	47	50					
46	3	8	14	21	27	34	41	48	50					
47	3	8	14	21	28	35	42	49	50					
48	3	8	15	21	28	36	43	50						
49	3	9	15	22	29	36	44	50						
50	3	9	15	22	30	37	45	50						

Table 49 CRITICAL VALUES FOR A NONPARAMETRIC TEST OF DISPERSION (Description: p. 151)

α = 0.05 critical values

n	2	3	4	5	6	7	8	9	10	11	12	13	14	15
6	-	3	4	6	7	9	11	13	14	16	18	19	21	23
7	-	3	5	6	8	10	12	14	16	18	20	22	23	25
8	-	3	5	7	9	11	13	15	18	20	22	24	26	28
9	-	3	6	8	10	12	15	17	19	22	24	26	28	31
10	2	4	6	9	11	13	16	18	21	23	26	28	31	34
11	2	4	7	9	12	15	17	20	22	25	28	31	34	36
12	2	4	7	10	13	16	19	21	24	27	30	33	36	39
13	2	5	8	11	14	17	20	23	26	29	32	35	39	42
14	2	5	8	11	15	18	21	24	28	31	34	38	41	45
15	2	5	9	12	16	19	23	26	30	33	37	40	44	47
16	2	6	9	13	17	20	24	28	31	35	39	43	46	50
17	3	6	10	13	17	21	25	29	33	37	41	45	49	50
18	3	6	10	14	18	22	26	31	35	39	43	47	50	
19	3	7	11	15	19	23	28	32	36	41	45	50		
20	3	7	11	16	20	25	29	34	38	43	47	50		
21	3	7	12	16	21	26	30	35	40	45	50			
22	3	8	12	17	22	27	32	37	42	47	50			
23	4	8	13	18	23	28	33	38	43	49	50			
24	4	8	13	18	24	29	34	40	45	50				
25	4	9	14	19	25	30	36	41	47	50				
26	4	9	14	20	26	31	37	43	49	50				
27	4	9	15	21	26	32	38	44	50					
28	4	10	15	21	27	34	40	46	50					
29	4	10	16	22	28	35	41	47	50					
30	5	10	16	23	29	36	42	49	50					
31	5	11	17	23	30	37	44	50						
32	5	11	17	24	31	38	45	50						
33	5	11	18	25	32	39	46	50						
34	5	12	19	26	33	40	48	50						
35	5	12	19	26	34	41	49	50						
36	6	12	20	27	35	42	50							
37	6	13	20	28	36	44	50							
38	6	13	21	29	37	45	50							
39	6	13	21	29	37	46	50							
40	6	14	22	30	38	47	50							
41	6	14	22	31	39	48	50							
42	7	14	23	31	40	49	50							
43	7	15	23	32	41	50								
44	7	15	24	33	42	50								
45	7	15	24	34	43	50								
46	7	16	25	34	44	50								
47	7	16	25	35	45	50								
48	7	16	26	36	46	50								
49	8	17	26	36	47	50								
50	8	17	27	37	48	50								

.ble 50 CRITICAL VALUES FOR THE PRODUCT MOMENT CORRELATION COEFFICIENT ($\rho = 0$)
(Table: p. 155)

Let (X_i, Y_i), $i = 1, \ldots, n$, be a sample from a bivariate normal population with true correlation $\rho = 0$. Let

$$r = \frac{\sum_{i=1}^{n}(X_i - \bar{X})(Y_i - \bar{Y})}{\left[\sum_{i=1}^{n}(X_i - \bar{X})^2 \sum_{i=1}^{n}(Y_i - \bar{Y})^2\right]^{1/2}}$$

Then $P(r \leq \text{tabulated value}) = \gamma$. Under the usual assumptions of simple linear regression, a test of $\rho = 0$ is a test that the slope of the true regression line equals zero.

Table 50 CRITICAL VALUES FOR THE PRODUCT MOMENT CORRELATION COEFFICIENT ($\rho = 0$)
(Description: p. 154)

	cumulative probability, γ					
n	0.750	0.900	0.950	0.975	0.990	0.995
3	0.7071	0.9511	0.9877	0.9969	0.9995	0.9999
4	0.5000	0.8000	0.9000	0.9500	0.9800	0.9900
5	0.4040	0.6871	0.8054	0.8783	0.9343	0.9587
6	0.3473	0.6084	0.7293	0.8114	0.8822	0.9172
7	0.3091	0.5509	0.6694	0.7545	0.8329	0.8745
8	0.2811	0.5067	0.6215	0.7067	0.7887	0.8343
9	0.2596	0.4716	0.5822	0.6664	0.7498	0.7977
10	0.2423	0.4428	0.5494	0.6319	0.7155	0.7646
11	0.2281	0.4187	0.5214	0.6021	0.6851	0.7348
12	0.2161	0.3981	0.4973	0.5760	0.6581	0.7079
13	0.2058	0.3802	0.4762	0.5529	0.6339	0.6835
14	0.1968	0.3646	0.4575	0.5324	0.6120	0.6614
15	0.1890	0.3507	0.4409	0.5140	0.5923	0.6411
16	0.1820	0.3383	0.4259	0.4973	0.5742	0.6226
17	0.1757	0.3271	0.4124	0.4821	0.5577	0.6055
18	0.1700	0.3170	0.4000	0.4683	0.5425	0.5897
19	0.1649	0.3077	0.3887	0.4555	0.5285	0.5751
20	0.1602	0.2992	0.3783	0.4438	0.5155	0.5614
21	0.1558	0.2914	0.3687	0.4329	0.5034	0.5487
22	0.1518	0.2841	0.3598	0.4227	0.4921	0.5368
23	0.1481	0.2774	0.3515	0.4132	0.4815	0.5256
24	0.1447	0.2711	0.3438	0.4044	0.4716	0.5151
25	0.1415	0.2653	0.3365	0.3961	0.4622	0.5052
26	0.1384	0.2598	0.3297	0.3882	0.4534	0.4958
27	0.1356	0.2546	0.3233	0.3809	0.4451	0.4869
28	0.1330	0.2497	0.3172	0.3739	0.4372	0.4785
29	0.1305	0.2451	0.3115	0.3673	0.4297	0.4705
30	0.1281	0.2407	0.3061	0.3610	0.4226	0.4629
35	0.1179	0.2220	0.2826	0.3338	0.3916	0.4296
40	0.1098	0.2070	0.2638	0.3120	0.3665	0.4026
45	0.1032	0.1947	0.2483	0.2940	0.3457	0.3801
50	0.0976	0.1843	0.2353	0.2787	0.3281	0.3610
55	0.0929	0.1755	0.2241	0.2656	0.3129	0.3445
60	0.0888	0.1678	0.2144	0.2542	0.2997	0.3301
70	0.0820	0.1550	0.1982	0.2352	0.2776	0.3060
80	0.0765	0.1448	0.1852	0.2199	0.2597	0.2864
90	0.0720	0.1364	0.1745	0.2072	0.2449	0.2702
100	0.0682	0.1292	0.1654	0.1966	0.2324	0.2565

Table 51 CRITICAL VALUES FOR THE MULTIPLE CORRELATION COEFFICIENT (Table: p. 157)

Let the distribution of Y, given variables $X_1, \ldots, X_k$, be normal: $N(\alpha + \beta X_1 + \cdots + \beta_k X_k, \sigma^2)$. From n independent observations of (k + 1)-tuples $(Y, X_1, \ldots, X_k)$, let Z be the least squares regression of Y on $X_1, \ldots, X_k$. Let $R = R_{y \cdot 1, \ldots, k}$ be the sample correlation coefficient of Y and Z. The tabled entry for f = n - k - 1 is such that when $\beta_1 = \beta_2 = \cdots = \beta_k = 0$, then $P(R \leq \text{tabled value}) = 0.95$ (first line) or = 0.99 (second line). The critical values R' for significance level α are related to the α critical value of the F distribution (Table 6), with k and n - k - 1 degrees of freedom, by

$$R' = \left[\frac{kF}{(n - k - 1) + kF}\right]^{1/2}$$

Table 52 RANDOM NUMBERS (Table: pp. 158-162)

This table presents "independent random digits." These pseudo-random numbers were generated by a linear congruential pseudo-random number generator and tested in several ways for randomness.

Table 51 CRITICAL VALUES FOR THE MULTIPLE CORRELATION COEFFICIENT (Description: p. 156)

0.05 critical values (top)
0.01 critical values (bottom)

	number of predictor variables					number of predictor variables			
f	2	3	4	5	f	2	3	4	5
1	0.999	0.999	0.999	1.000	18	0.532	0.587	0.628	0.660
	1.000	1.000	1.000	1.000		0.633	0.678	0.710	0.736
2	0.975	0.983	0.987	0.990	19	0.520	0.575	0.615	0.647
	0.995	0.997	0.997	0.998		0.620	0.665	0.697	0.723
3	0.930	0.950	0.961	0.968	20	0.509	0.563	0.604	0.636
	0.977	0.983	0.987	0.990		0.607	0.652	0.685	0.712
4	0.881	0.912	0.930	0.942	21	0.498	0.552	0.593	0.624
	0.949	0.962	0.970	0.975		0.596	0.641	0.674	0.700
5	0.836	0.874	0.898	0.914	22	0.488	0.542	0.582	0.614
	0.917	0.937	0.949	0.957		0.585	0.630	0.663	0.690
6	0.795	0.839	0.867	0.886	23	0.479	0.532	0.572	0.604
	0.886	0.911	0.927	0.938		0.574	0.619	0.653	0.679
7	0.758	0.807	0.838	0.860	24	0.470	0.523	0.562	0.594
	0.855	0.885	0.904	0.918		0.565	0.609	0.643	0.669
8	0.726	0.777	0.811	0.835	25	0.462	0.514	0.553	0.585
	0.827	0.860	0.882	0.898		0.555	0.600	0.633	0.660
9	0.697	0.750	0.786	0.812	26	0.454	0.506	0.545	0.576
	0.800	0.837	0.861	0.878		0.546	0.590	0.624	0.651
10	0.671	0.726	0.763	0.790	27	0.446	0.498	0.536	0.568
	0.776	0.814	0.840	0.859		0.538	0.582	0.615	0.642
11	0.648	0.703	0.741	0.770	28	0.439	0.490	0.529	0.560
	0.753	0.793	0.821	0.841		0.529	0.573	0.607	0.633
12	0.627	0.683	0.722	0.751	29	0.432	0.483	0.521	0.552
	0.732	0.773	0.802	0.824		0.522	0.565	0.598	0.625
13	0.608	0.664	0.703	0.733	30	0.425	0.476	0.514	0.545
	0.712	0.755	0.785	0.807		0.514	0.558	0.591	0.618
14	0.590	0.646	0.686	0.717	40	0.373	0.419	0.455	0.484
	0.694	0.737	0.768	0.791		0.454	0.494	0.526	0.552
15	0.574	0.630	0.670	0.701	60	0.308	0.348	0.380	0.406
	0.677	0.721	0.752	0.776		0.377	0.414	0.442	0.467
16	0.559	0.615	0.655	0.687	90	0.254	0.288	0.315	0.338
	0.662	0.706	0.738	0.762		0.312	0.343	0.368	0.390
17	0.545	0.601	0.641	0.673	120	0.221	0.251	0.275	0.295
	0.647	0.691	0.724	0.749		0.272	0.300	0.322	0.342

Table 52 RANDOM NUMBERS (Description: p. 156)

7972	8825	4491	4617	1295	3069	9319	5291	0521	7748
1445	2793	9608	8735	9858	2584	9193	3207	2178	0272
2593	5218	5646	7715	2283	8846	0375	9367	8670	2107
5583	7102	6697	7251	9566	1091	6589	0194	4114	3734
9189	9058	1819	5648	7382	7031	6524	9000	1759	8206
7114	1643	0903	1671	4848	4232	5425	9602	6624	5355
5666	6219	9104	0202	9566	3688	8501	6640	6587	0821
1720	2534	4486	4224	7269	5178	1263	4783	0027	9073
4515	1814	2274	1899	1161	1435	2885	5605	9548	4022
6102	2716	3093	2525	6606	7458	1980	9323	3793	2882
6687	7556	4224	5780	5898	6534	4803	0681	2717	3388
3645	4892	3068	7422	2535	5328	9686	0986	4963	2123
7721	8186	4010	4660	7247	2264	0367	2258	5135	1646
0340	5096	1998	4578	9015	8880	8614	9285	7010	1354
7244	0025	2667	0863	1598	1315	0686	8416	9554	2313
4583	2194	3662	7370	4653	6464	2528	1733	6576	9891
6421	3955	4831	9379	4997	7185	8480	4012	6041	9867
9198	4597	9831	8281	9200	1376	4211	7883	2634	4464
5266	2935	7831	7996	2259	3093	2302	2435	0114	4341
8641	4464	8244	9537	8464	8200	1982	0665	8838	9783
4579	8486	5534	4698	4198	2185	9833	1519	5607	9233
2978	7578	1008	4239	7306	2737	8090	2443	0353	6964
8513	5476	0860	1566	7774	4690	6526	6997	5253	7417
2357	6183	7206	3511	9337	4943	4100	6195	7754	4879
9862	2944	3886	1137	2378	2393	8196	2085	6383	8908
4025	3236	0812	8115	8520	3494	5238	1613	4317	8828
1006	2738	4535	1699	8618	6251	7085	9632	5378	5429
8361	6555	0276	0124	1676	2640	2376	1616	5699	2069
9497	5887	8776	7742	8510	8978	4010	5337	8194	6804
5223	9191	4820	0907	9470	6529	3015	3013	1876	3163
0079	2555	4862	0955	3157	8161	9626	3183	6389	4595
2193	4146	3649	8367	7824	3697	0805	0032	5629	2684
0651	3531	5444	1305	7703	6332	1660	7062	5585	8166
7109	1834	5520	3582	2525	1457	3972	9453	3245	9156
7815	8210	3996	5534	2218	0846	4468	4372	7175	7895
3664	7531	5107	6122	6487	1536	5244	3110	5293	8616
9725	4917	9161	8922	4191	2916	4859	2169	2830	9178
6619	4956	1942	5087	3487	8622	0492	7614	4975	6174
6629	3646	4313	6487	8059	9746	4410	3194	3539	6821
4668	4548	9724	8655	1350	6737	1290	3938	7051	0660

Table 52 RANDOM NUMBERS (Description: p. 156)

7149	2544	3239	3762	1068	0395	6953	7618	1888	3167
2753	3391	9744	4673	1085	3721	3487	7909	4642	0467
5079	9572	5163	8077	0980	2678	8855	1565	0914	0834
3300	7047	1277	1149	8290	9991	3482	7433	7513	7980
0486	2504	1102	8121	3639	4383	0048	2551	4897	4891
8551	7843	4949	7750	9195	6581	5682	1273	4073	8796
6741	9012	8203	8618	8422	2263	6193	8822	9482	2597
9834	3936	1873	8113	7318	7062	0881	7875	4700	7363
6469	5700	3455	8539	7650	5646	3843	0270	0246	1740
3394	0628	3870	2752	5840	8952	3093	9911	1015	4763
6475	8078	8036	8754	9167	6392	3770	3293	5579	7627
8458	3448	3600	3124	2531	9915	6488	5655	1505	7537
9832	4487	0914	2558	6619	5461	9627	1067	1386	6764
7463	0882	7620	0361	8385	8843	7737	1430	5586	2111
4021	9947	6481	4983	0608	4150	9289	3287	7633	4088
6679	0755	6957	6670	9133	9674	9446	2237	5834	2340
5671	7829	9437	2708	6356	2567	2291	6490	1880	6624
5234	7825	9605	7103	3998	5495	1852	2297	9796	4845
3501	5001	4344	4131	6700	0719	5337	4475	6450	7548
6878	4271	1356	7683	1614	4497	8039	8162	6941	6803
5904	6463	0012	4913	6720	6423	3476	9044	0450	1634
7560	7076	0472	3023	8006	3962	2679	2902	2827	2159
1157	5602	8827	2957	2539	1446	8092	0469	7879	3455
4743	4561	3299	5954	2229	5673	9572	9276	0123	2156
1999	5657	6780	7278	1752	1332	3165	1646	0941	5637
3888	8094	0727	0908	5289	1886	3807	2585	3170	4002
9153	1634	0076	4662	9131	8677	7252	6103	2441	2770
9328	8102	0569	9566	1843	7540	0622	5226	3799	3563
3936	4191	5810	5509	6567	7634	0354	4157	8350	5219
2256	1403	6280	7088	9138	0494	8730	0728	9201	1891
0725	8492	4389	1187	0556	8859	7058	2470	3578	9093
4206	3710	6920	3784	1901	9870	4966	9197	9428	7247
2474	5815	8607	6061	3447	4024	4216	5463	4683	3596
2661	4146	7711	3531	0583	5691	6270	5613	7682	7166
4619	7132	8083	6895	4203	7682	5649	3397	8121	2677
5780	4475	8359	5681	6642	2174	5033	8020	1719	6853
2977	4489	0623	0135	7718	0395	0120	5822	2008	4237
7055	6996	6432	3740	1329	6996	1172	1315	1142	2368
2422	9560	3737	8918	2502	1017	1569	5120	7456	3184
4926	6067	5093	6786	9784	4476	4004	3163	0606	5078

Table 52 RANDOM NUMBERS (Description: p. 156)

3248	2128	0534	3358	5026	2216	6523	6301	0926	7040
5888	1060	2375	5166	8760	5266	8212	5209	7637	7108
2984	3806	6233	9184	9247	3694	2022	8048	4087	8331
4101	9737	2089	0566	2337	3460	5970	8827	9971	3539
7042	2076	1965	3902	1960	5761	4964	9525	0766	4636

7272	5624	4725	2362	3740	0602	8817	0602	3065	9549
1955	8119	5032	3097	5829	5748	5832	9690	9271	6093
4763	9758	6749	8700	2871	3261	3784	2656	5638	3285
2454	1417	7774	2968	9278	5971	9944	3830	6103	2998
8872	0067	8097	9074	9777	0356	5956	1172	5448	7247

4437	1990	4306	0275	1295	1411	3715	6347	4825	1136
5262	3443	9582	1139	1411	2227	1719	6021	7034	2046
7401	3789	0050	6900	0800	3125	0425	4902	8667	1936
6242	1519	1846	5038	5169	0389	7917	5317	0165	5447
7782	7602	0151	3797	5459	8980	3500	0801	4559	0801

4407	3458	6100	6544	4608	0557	3482	6759	2474	1144
6242	0529	3733	5993	8688	4503	1250	3488	8275	4896
5228	6849	7652	4340	2486	8216	5433	6169	8420	0286
2974	1807	2432	9534	7408	7681	4204	8114	9222	3356
9011	4259	6107	3138	0219	3678	6258	2785	5167	3751

8286	3346	4203	2942	8934	6659	8563	1318	3732	0064
0579	7660	0372	3120	7347	5482	5894	0289	5966	4061
3518	6212	6755	8646	1523	4162	0868	0942	2957	4092
3087	5694	2439	6956	0779	8043	4816	7827	0782	6380
3855	4962	3944	6233	2943	0255	2300	5611	4475	9755

1171	2606	0758	2546	7675	7693	5805	8899	0187	0214
9072	4601	9310	4407	4000	8123	9732	9430	1886	3252
4429	0521	0979	3282	2035	9592	2084	8293	2929	3308
8620	9477	6129	6136	5930	3727	7167	5418	3333	8637
2266	8860	2810	6911	8242	3467	9410	2145	4891	8403

1042	6133	4804	5310	9383	0292	1249	2836	6083	0630
2393	4941	1756	4544	5207	1940	0742	5152	1531	8470
0215	9674	1613	8090	4651	4220	2274	7983	0835	3201
0125	0358	1872	3767	4384	6853	2395	1201	4845	0352
7248	1665	6909	2639	9042	8553	2814	2337	2183	2929

1165	9895	9938	5309	2405	4191	2893	4826	8481	1243
3664	6818	1932	0575	3595	7622	9187	7530	5392	9495
7050	6338	9947	5676	7845	2860	5896	6604	1147	2683
4749	9537	3366	2142	5882	2918	0040	1562	8632	9007
9283	8645	1105	5256	2718	2250	3578	2827	2793	4768

Table 52 RANDOM NUMBERS (Description: p. 156)

3674	0375	2143	5283	7489	7971	1610	8832	5597	2183
7495	6197	0621	9374	2302	3531	3154	2801	7827	3381
7369	3161	0058	4587	0556	4712	7754	9893	5019	3840
9298	0715	9527	7126	6631	1385	6162	1749	3402	4933
1451	0489	4155	8058	6305	7427	1493	5357	6151	1366
0034	5065	4460	2597	5094	2246	2486	8603	1247	8581
3318	4999	3091	9791	8241	4149	1694	9932	0173	1587
5618	5451	9747	6701	7517	4873	5990	7206	4025	3991
0163	3791	1079	0894	3043	9492	9599	6923	4811	1209
7594	1723	1861	9341	0589	6192	5114	7630	2638	3431
2911	1483	3640	5795	2874	2426	0967	6821	1973	4187
7487	1559	8588	3283	6030	4254	8508	0244	5950	3023
8112	7808	0085	4796	5858	4506	3942	6317	3652	9078
5833	7962	9755	7847	1569	9952	6020	4862	8515	0229
0554	2470	4978	3520	5603	4321	0505	0355	4312	7253
9467	4736	1894	1816	5017	0900	7685	2672	7588	2434
8936	5661	2595	7923	3911	2155	4889	0600	5602	7298
4848	0447	5000	6262	9885	5293	5727	1204	5682	3484
5593	6615	4528	6530	5948	8424	6778	4431	4542	8135
6793	0951	2362	0697	9115	5610	9889	1606	5697	4968
3009	3822	3054	0681	1840	0142	6416	9954	3138	9238
7760	8030	5620	1909	5310	3742	5050	1018	7556	1788
8836	6275	5857	3167	2229	7464	0988	7424	2746	9588
0590	4114	4308	8231	9532	4240	3250	8207	9040	2813
3123	1703	0137	8490	9999	7234	6241	8431	6229	4897
2751	5059	5835	0200	7188	6268	1915	1461	9195	6641
6991	8109	2677	5756	3019	7424	8580	8613	4421	5046
4085	6467	0082	3997	9200	8764	2601	7086	3690	6115
1044	3349	7679	0732	8958	5518	8853	3655	4638	6796
1132	9469	5731	5848	6325	1862	5419	2980	2485	5997
8959	4667	5618	7985	7122	8207	7258	3447	7142	6565
8765	2063	7612	2673	5599	0006	3373	6386	2472	9514
4767	7846	0218	7398	5555	3002	6844	7951	2717	0915
8447	0660	2832	2548	5238	9785	9821	1931	0579	7710
9711	9159	7854	9691	6211	5255	1718	0478	3498	6123
2121	8444	7609	8000	0711	7415	0541	7879	7610	8219
1499	4819	8052	7931	8670	0986	0144	6045	1387	7379
1371	4080	2696	9055	4345	3989	3449	6305	9626	5037
4401	9018	5200	4532	6452	4466	6929	2217	0007	0759
3943	6036	5656	9960	7300	0287	1640	6281	3522	2162

Table 52 RANDOM NUMBERS (Description: p. 156)

9488	0436	8929	8664	8084	3241	7993	8604	4302	5062
6316	5180	7933	7485	1859	1974	6025	3348	6751	9516
1967	2256	1058	3346	0647	2261	5375	7329	1201	0789
2740	6462	2347	2862	4252	6107	1941	0129	9112	5977
4437	7513	6894	0879	9503	2654	6805	0705	1540	5526
1669	4217	7655	1722	7278	4852	1050	7336	1453	0431
0898	5073	2246	4146	7891	9638	1772	0465	8941	8027
2561	1936	5710	0248	1698	2683	3898	3718	5097	9621
5695	3825	4020	7421	3641	3017	1872	6564	1874	5476
8090	2246	8454	9841	3152	4405	9328	5233	3080	3726
3494	0464	0237	0880	1547	9557	4215	5809	4297	6144
2645	2733	3976	3560	0221	3248	5222	1330	6948	3940
2297	0798	1604	7062	1168	6690	0668	1649	1926	5944
5182	1497	0775	1517	7738	6008	7118	7636	1835	3626
7907	3151	2135	3407	8663	9912	2442	5064	7776	1077
8883	0208	0597	7680	6355	0278	2337	0500	7150	9153
9183	3980	2658	8757	5175	6784	7334	5796	6820	9175
6744	2430	5308	0928	9756	8952	6973	5477	6189	8178
3503	7529	0086	1727	9293	1430	5560	5982	5450	6077
8937	1138	6208	3729	7694	8015	1891	8264	5252	9344
3072	0071	3261	5532	3606	1941	8716	0766	1374	5244
6552	6978	8890	3982	7746	0814	3453	5337	3507	7771
6616	8401	7398	3070	7578	0469	5400	6558	2330	3445
5760	5260	3941	3823	6765	7218	5291	0410	2594	3127
7085	4711	6928	4653	5054	3735	4240	3757	7604	4788
5684	9754	8301	1979	0529	0169	6292	7518	0573	6764
4872	5624	2377	5531	0826	4729	1543	1877	6569	2732
7497	3243	8484	7287	0705	7500	0455	2185	5986	2745
5005	2072	5190	7399	6715	8980	0011	4602	8858	1244
7181	2852	0625	7191	0170	5646	7225	6166	0993	3708
0600	1295	6604	3502	9779	7038	3181	8028	6311	4522
4063	3376	1973	4407	8422	5814	5256	5631	2069	1087
7444	4966	0543	9019	5165	5242	3657	8992	1236	4233
0224	7107	1012	2189	4512	8302	4348	9364	0656	1414
8919	4243	5968	0648	1698	5161	7769	6487	8010	5511
0964	5503	7567	1861	4047	1824	4593	3377	6918	0689
5347	3528	3614	7454	4982	8800	5420	3833	8489	8673
4306	2057	1747	6436	3890	7049	7567	7741	8429	6617
9678	4827	8912	1447	3307	7338	2981	2585	9305	5573
3539	4194	2646	6149	5651	8856	3719	5547	4635	1481

REFERENCES TO THE TABLES

The reference numbers for each table are given in parentheses; C, Chapter; S, Section(s); P, beginning on page(s). (References are listed on pages 165-166.)

Table	References
1	(1)S2.7; (11)S3.4; (18)S1.2; (22); (26)C2
2	(11)S4.4, 6.2, 6.3; (18)S2.1; (22); (26)
3	(26)S2.14
4	(26)S4.13
5	(11)S3.3, 10.2; (13); (10)S3.1; (22); (24); (26)P104-111
6	(11)S4.4, 10.6; (18)S4.1; (22); (24); (26)
7	(2)P111-232
8	(18)S5.3
9	(12)S20.37-20.40; (18)S5.4
10	(19)
11	(18)S6.1; (22)
12	(9)P950, 967; (18)S6.2
13	(6); (18)S6.3
14	(18)S6.4
15	(18)S6.5; (22); (24)S3.6
16	(21)
17	(4)C4; (9)C21, P967
18	(9)S2.1, P973; (12)S34.10-34.22; (27)P2.39-2.42
19	(8)P313-318, 542
20	(4)S10.3; (9)C10
21	(5)P17
22	(1)S2.6; (11)S3.2; (12)S5.8-5.12; (18)S9.3
23	(18)S9.4; (22)
24	(1)S2.7; (11)S3.1; (12)S5.2-5.7; (18)S9.5; (26)C8
25	(18)S9.6; (22)
26	(18)S10.1

Table	References
27	(18)S10.3
28	(3)S5.8; (10)S4.1; (14)S1.2, 2.3; (18)S11.4; (25)P116-127
29	(3)S7.2-7.3; (10); (14); (18)S12.1; (25)P68-75
30	(14); (18)12.2; (25)P68-75
31	(3)S11.2-11.3; (14)S7.5C; (18)S12.5; (25)P136-145
32	(17); (18)S12.12
33	(3)S13.1; (10)C8; (22); (25)P213-223
34	(3)S5.13; (10)P191-192; (14)S7.3; (22); (25)P202-213
35	(10)S1.1; (14)S5.2; (25)P184-193; (26)C8
36	(10)S1.2; (14)S5.5B
37	(10)S6.3
38	(15); (16)
39	(3)S5.12; (10)S7.1; (14)S6.2; (25)P166-172
40	(10)S7.2; (20)
41	(10)S7.3
42	(3)S13.5; (18)S15.1; (25)P47-52
43	(3)S13.3; (10)S10.1; (25)P127-136
44	(3)S13.3; (10)S10.1; (14)S1.6; (25)P127-136
45	(3)S13.3; (14)S1.6; (18)S15.6; (25)P131, 279
46	(3)S8.2; (18)S18.2; (25)P96-104
47	(18)S18.3
48	(18)S18.4; (28)
49	(14)S1.6; (18)S18.5
50	(7); (18)S19.1
51	(7); (18)S19.3

REFERENCES

1 P. Armitage, Statistical Methods in Medical Research, Wiley, New York, 1971.

2 K. O. Bowman and M. A. Kastenbaum, Sample Size Requirement: Single and Double Classification Experiments, in Selected Tables in Mathematical Statistics, Vol. 3 (H. L. Harter and D. B. Owen, Eds.), American Math. Soc., Providence, R.I., 1975.

3 J. V. Bradley, Distribution-Free Statistical Tests, Prentice-Hall, Englewood Cliffs, N.J., 1968.

4 I. W. Burr, Statistical Quality Control Methods, Marcel Dekker, New York, 1976.

5 J. M. Cameron, Quality Progress, 17-19 (September, 1974).

6 J. F. Daly, Ann. Math. Statist., 17: 71 (1946).

7 A. P. Dempster, Continuous Multivariate Analysis, Addison-Wesley, Reading, MA., 1969.

8 W. J. Dixon and F. J. Massey, Jr., Introduction to Statistical Analysis, McGraw-Hill, New York, 1951.

9 A. J. Duncan, Quality Control and Industrial Statistics (3rd ed.), Richard D. Irwin, Inc., Homewood, Ill., 1968.

10 M. Hollander and D. A. Wolfe, Nonparametric Statistical Methods, Wiley, New York, 1973.

11 R. V. Hogg and A. T. Craig, Introduction to Mathematical Statistics (3rd ed.), Macmillan Co., New York, 1970.

12 M. G. Kendall and A. Stuart, The Advanced Theory of Statistics, Vols. 1, 2, 3 (3rd ed.), Hafner, New York, 1969, 1973, 1976.

13 H. O. Lancaster, The Chi-squared Distribution, Wiley, New York, 1969.

14 E. L. Lehman, Nonparametrics: Statistical Methods Based on Ranks, Holden-Day, San Francisco, 1975.

15 B. J. McDonald and W. A. Thompson, Jr., Biometrika, 54, 487 (1967).

16 R. E. Odeh, Technometrics, 9, 271 (1967).

17 P. S. Olmstead and J. W. Tukey, Ann. Math. Statist., 18, 495 (1947).

18 D. B. Owen, Handbook of Statistical Tables, Addison-Wesley, Reading, MA., 1962.

D. B. Owen and W. F. Frawley, J. Quality Technology, 3: 69 (1971).

20 E. B. Page, J. Am. Statist. Assn., 58: 216 (1963).

21 E. S. Pearson, Biometrika, 39: 130 (1952).

22 E. S. Pearson and H. O. Hartley, Biometrika Tables for Statisticians, Vol. 1 (3rd ed.), University Press, Cambridge, 1970.

23 C. R. Rao, Linear Statistical Inference and Its Applications, Wiley, New York, 1965.

24 H. Scheffé, The Analysis of Variance, Wiley, New York, 1959.

25 S. Siegel, Nonparametric Statistics for the Behavioral Sciences, McGraw-Hill, New York, 1956.

26 G. W. Snedecor and W. G. Cochran, Statistical Methods (6th ed.), Iowa State University Press, Ames, IA., 1967.

27 Statistical Research Group, Columbia University, Sequential Analysis of Statistical Data, Columbia University Press, New York, 1945.

28 S. S. Wilks, Ann. Math. Statist., 13: 400 (1942).